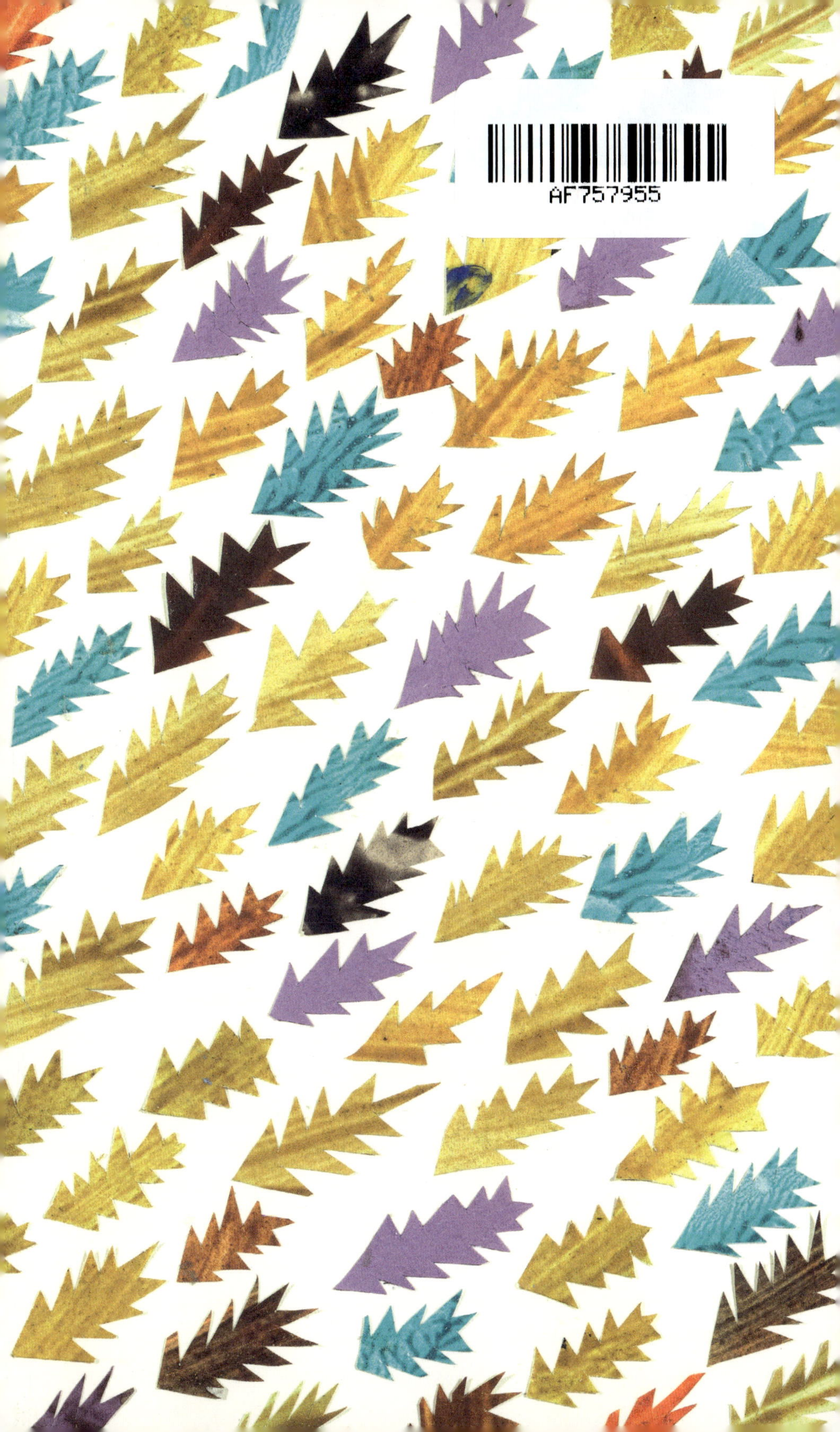

DRAWN TO NATURE

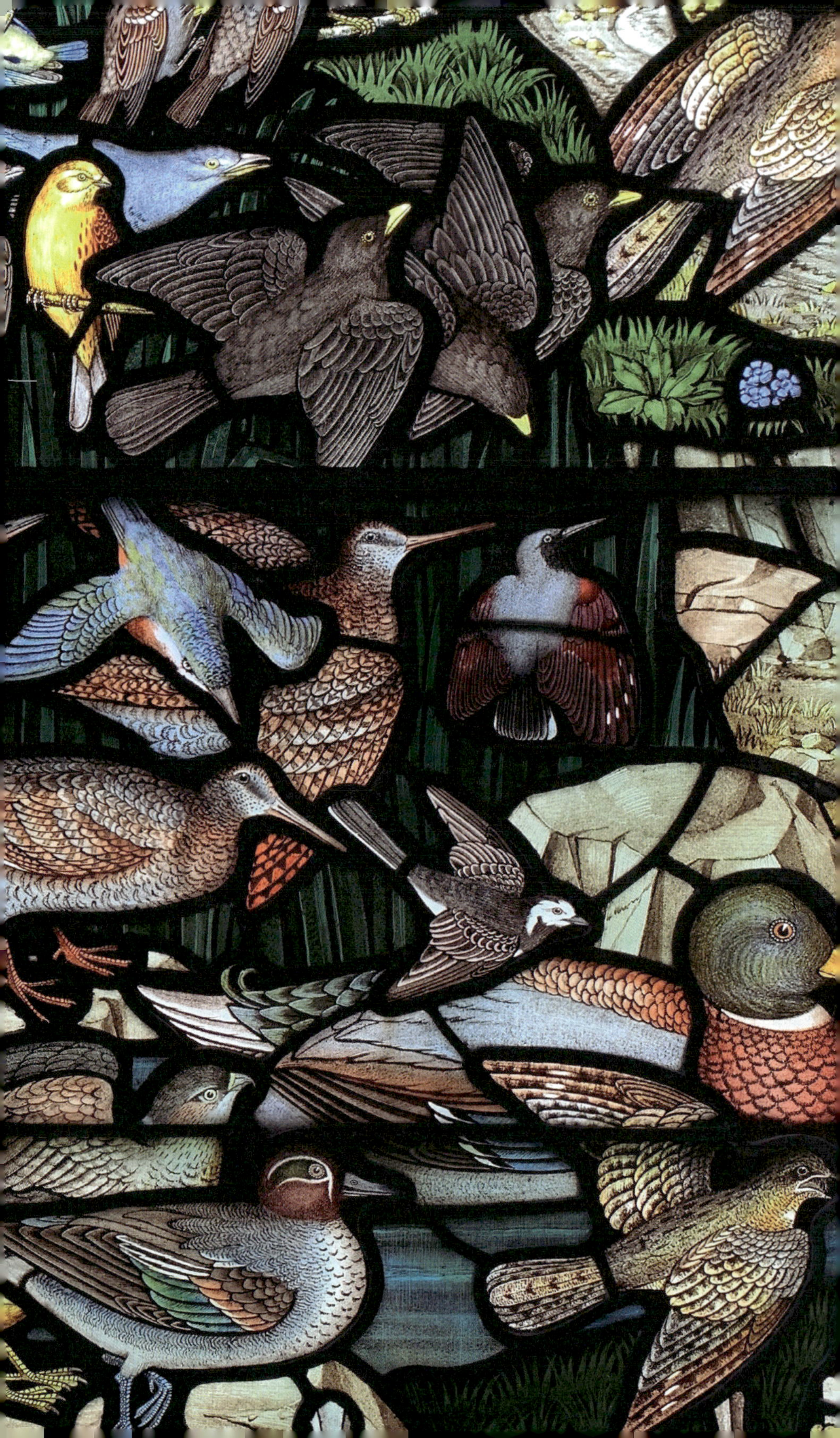

DRAWN TO NATURE

Gilbert White and the Artists

SIMON MARTIN

With essays by
SIR DAVID ATTENBOROUGH
and VIRGINIA WOOLF

Pallant House Gallery

Distributed by
Yale University Press

First published in the United Kingdom in 2021 by
Pallant House Gallery

Please note: Gilbert White's original spellings have been used when quoting from his *Natural History*. These occasionally differ from modern English spellings.

Pallant House Gallery
8–9 North Pallant
Chichester
West Sussex
PO19 1TJ
Tel: +44(0)1243 7744557
info@pallanthousegallery.co.uk
www.pallant.org.uk

British Library Cataloguing-in-Publications Data
A catalogue record for this book is available from the British Library

ISBN 978-1-869827-75-5

Library of Congress Control Number: 2021941745

Design by Adrian Hunt
Set in Berthold Akzidenz-Grotesk
Printed and bound in the UK by Gomer Press

Distributed for Pallant House Gallery by Yale University Press,
New Haven and London

(frontispiece) Detail from *St Francis Preaching to the Birds*, 1920 (see fig.29).

FSC
www.fsc.org
MIX
Paper from responsible sources
FSC® C114687

Contents

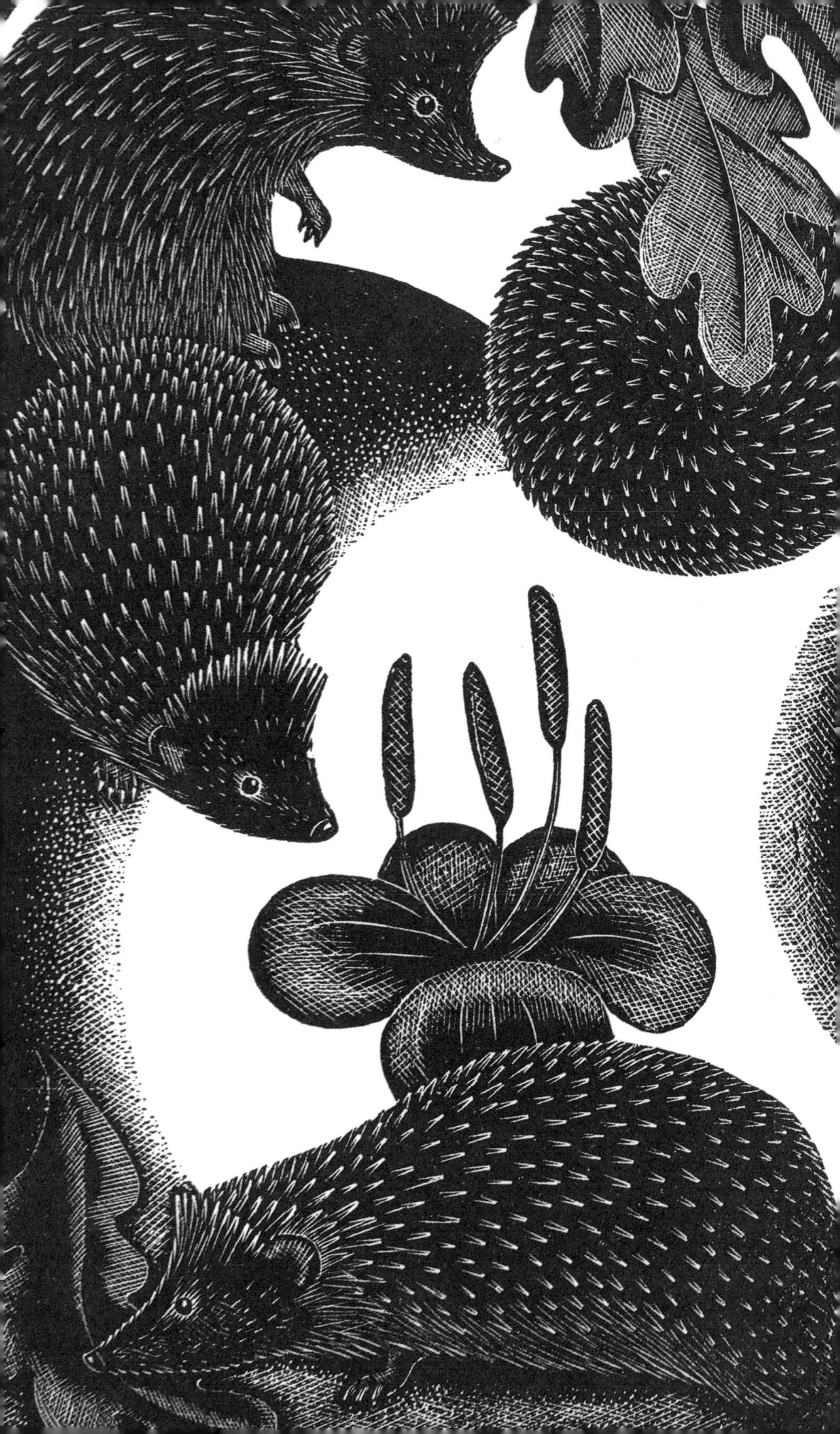

FOREWORD

Simon Martin
Director, Pallant House Gallery
August 2021

Since its publication over 230 years ago, Gilbert White's *The Natural History and Antiquities of Selborne* has inspired generations of naturalists, writers and artists. It is one of the most-loved books in the English language, and has never been out of print.

It is surprising that, despite the multitude of books on White, none of these has explored his remarkable artistic legacy: the many beautifully illustrated editions of his *Natural History*. It is a rich history involving celebrated Modern British artists, amongst them Gertrude Hermes, John Nash, John Piper and Eric Ravilious, but also forgotten figures such as Claire Oldham.

In 2020, the exhibition *Drawn to Nature: Gilbert White and the Artists* at Pallant House Gallery brought together many of these illustrations to celebrate the tercentenary of the birth of the 'first naturalist'. This book has grown from that exhibition. Written during the covid-19 pandemic, when many people found themselves looking to nature outside their windows, it takes us from the first edition in 1789 to the present day.

With over 400 editions and reissues of the *Natural History* published in Britain and internationally, this book can make no claim to comprehensiveness, but it will hopefully serve as a testament to White's influence and a reminder of the importance of both observing and caring for the natural world. White wrote about the interconnectedness of natural life and encouraged us to look and to take note of the flora and fauna around us. In a planetary emergency, his writings remain as important as ever.

FIG. 1
Claire Oldham, *Hedgehog*, 1946.
Wood-engraving on paper.
Illustration to Gilbert White,
The Natural History of Selborne
James Fisher, ed., London:
Cresset Press, 1947.

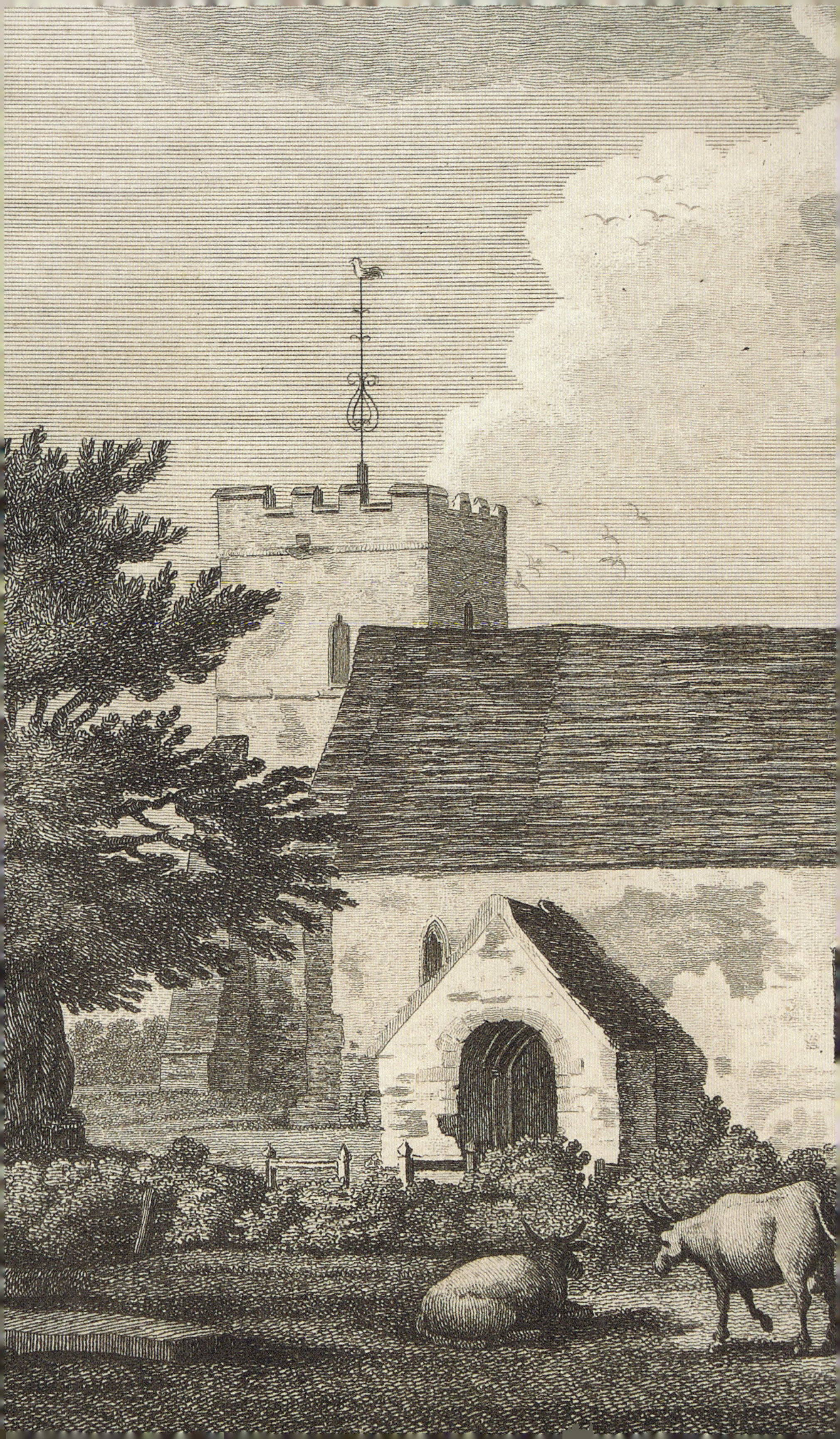

1.
INTRODUCTION

Sir David Attenborough

First published in *The Gilbert White Museum Edition of The Natural History of Selborne by Gilbert White*, London: Shepheard-Walwyn, 1977.

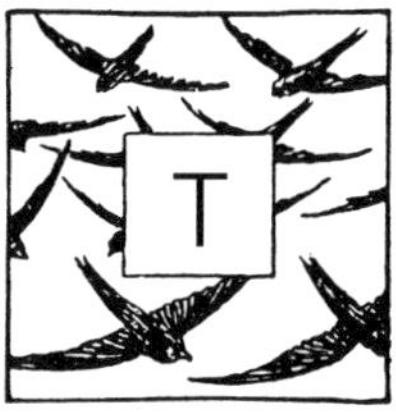

The village of Selborne lies in Hampshire, surrounded by fields that have been farmed since prehistory and overlooked by chalk hills crowned with beeches. It is small and pretty and not dramatically different from a dozen others in this part of southern England. Yet its name is known throughout the world, and its animals and plants are familiar to people in Italy and Sweden, in Australia and the United States, in Germany and Japan. It owes its fame entirely to this one book, written by the man who served as its curate two hundred years ago – a book that, to judge from the number and variety of editions in which it has appeared, is one of the most continuously popular books in the English language.

Gilbert White was born in 1720 in the vicarage alongside the church, where his grandfather was vicar and where his parents were visiting at the time of his birth. When he was nine, his family came to live permanently in Selborne. As a boy, he played on the Plestor, the village green, beside the churchyard wall. He went to school in Basingstoke and then went on

FIG.2
Detail from *South View of Selborne Church* (see fig.17).

to Oxford University, where he became a Fellow of Oriel College. But his ties remained with Selborne, and when he was 31 he became curate there. As the years passed he served in neighbouring parishes, but his last post was back in Selborne. He repeatedly refused promotion in order to remain living there. He never married. When he was 69, he published this, his one and only book. Four years later, at the age of 73, he died and was buried, as requested in his will, on the unfashionable northern side of the church, within a few yards of where he was born. It would be difficult to find a less eventful life, in a worldly sense.

But his days, for him, were packed with incident. He spent a great deal of his time in the garden of 'The Wakes', the family house in the village of which he eventually became the owner. Nothing was too small or humble to escape his investigations. He was fascinated by the earthworms that lived beneath the lawn. He came out at night with a candle and saw them lying out of their burrow in copulation and observed, rather tartly, that they are 'very venerous and seem to engender all the year'. He examined their casts, deduced that they are earth, and pointed out that gardeners and farmers who disliked them would soon find their land 'cold, hardbound and void of fermentation and consequently sterile' without them. 'These hints,' he added, 'we think proper to throw out in order to set the inquisitive and discerning to work'. It was 104 years before anyone was sufficiently inquisitive and discerning to take up these hints – and then it was no less a person than Charles Darwin. He wrote a monumental book on the subject and it acknowledged his debt to White.

White was never worried about appearing undignified in his investigations or even, indeed, a little dotty. He wondered whether bees could hear, so he went out to a hive and shouted at them through a large speaking trumpet 'with such an exertion of voice as would have hailed a ship at the distance of a mile and still these insects pursued their various employments undisturbed and without showing the least sensibility or resentment'. He had a musical friend who, doubtless egged on by White, tested all the owls in his neighbourhood with a pitch pipe and found that they all hooted in B flat. And that subject had to wait until the age of tape recorders before other people realised how fascinating it is and what wide-ranging propositions can be deduced from it.

Having made such observations as these, he meticulously noted them down in his journal. One can imagine him sitting in his study

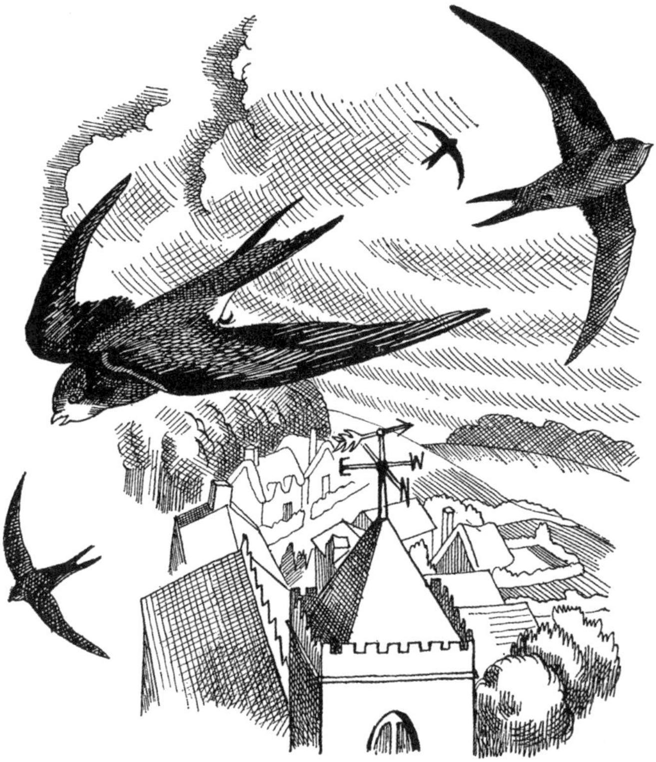

FIG.3
John Nash R.A., '*Dash round the steeple, unsubdu'd of wing*', 1951. Line drawing. Illustration to letter 1 to Thomas Pennant, in Gilbert White, *The Natural History of Selborne with Drawings by John Nash*, John Lewis, ed., London: Lutterworth Press, London, 1951.

in The Wakes, a wig on his head, a quill in his hand, wearing the black clothes of a clergyman. But we have no portraits of him, except for two small sketches in a copy of Pope's translations of the Iliad which he was given as a prize, by Pope himself, when he was at Oxford. From the letter and accounts of others, it seems that he was quiet, perhaps a little prim, and certainly very gentle and even-tempered. It is almost impossible to imagine him getting angry. Indeed, the only account we have of him even approaching irritation concerns his faithful servant, Thomas, who came to him to confess that he had broken a glass. 'Broken a glass,' said White, 'How did you do that?' 'I'll show you sire,' said Thomas and promptly picked up another wine glass and dropped it on the floor. And White's only response was to say, 'There, go along Thomas, you are a great fool,' and adding under his breath, 'And I was as great a one for asking such a foolish question'.

But if White had been just a quaint, slightly eccentric eighteenth-century curate, it is difficult to believe that his one book would have remained of such continuing fascination. Why should this be so?

He certainly made several original scientific discoveries. For many years he watched, with close attention, a small brown bird that visited his garden every summer. His fellow naturalists called it a 'willow wren'. But White was puzzled by it. It seemed to have three quite different songs. And then with that penetrating power of observation which was one of his most remarkable skills, he saw – and was the first person to do so, it seems – that there were three different species that had looked, to every eye except his, identical. We now know them as the chiff-chaff, the willow warbler and the wood warbler. And the easiest way to tell them apart is still their song.

In the surrounding cornfields, he noticed a little mouse that was quite different from the house mouse. It built beautiful nests, 'wonderful procreant cradles' White calls them, 'most artificially platted and composed of the blades of wheat, perfectly round and about the size of a cricket ball; with the aperture so ingeniously closed that there was no discovering to what part it belonged'. The little creature that made it, he correctly decided, was a hitherto unidentified species. It is Britain's smallest mammal, the harvest mouse.

He watched the swifts flying around the tower of his church. They are difficult birds to observe for they are truly swift, one of the fastest flying of all birds. But White detected, with total certainty, that they copulate on the wing – and he did so without the aid of binoculars. He was obsessed by these swifts and their relatives, swallows and martins. He refers to them over and over again in his letters; and from his daily journal, which was not published until 1931, 138 years after his death, we can see how much joy they brought him. In 1768, the very first year that he kept the journal, he noted on April 5th that the nightingale had arrived, and he gives the entry an exclamation mark of pleasure. But eight days later on, he writes 'Hirundo domestic' – the swallow – and that entry he awards three exclamation marks of sheer delight. One thing about the swallows remained a puzzle to him to the end of his life. Did they or did they not spend the winter in England in hibernation? He knew that some of them flew to Africa, but he found it very hard to believe that the young, only a few weeks old, could possibly fly all that way. Some people told him that swallows retired under water for the winter and

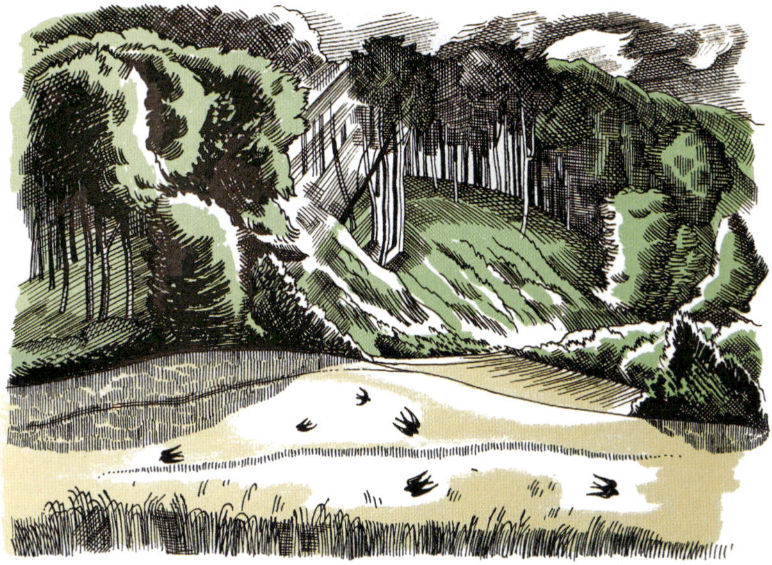

FIG.4
John Nash R.A., *Cornfield Below the Hanger*, 1951/1972. Colour lithograph on paper. Illustration to Gilbert White, *The Natural History of Selborne*, Ipswich: Limited Editions Club, 1972.

that martins burrowed their way underground. But White believed nothing that he did not see with his own eyes. So he searched in the thatch of the village cottages. He employed parishioners to explore the woods above his house. He looked in the river banks and the sandpits. But he never found what he called 'the secret dormitories', and a century after his death the subject was still being debated. Today we know that although at least one bird, the whip-poor-will of North America, does go into a form of hibernation, White's beloved swallows and martins do not.

However, even such important discoveries as White did make are hardly sufficient to account for the immortality of his book. I suspect that its enchantment comes from the beguiling picture it gives us of a man in total harmony with his world. He had two favourite walks around his village, the Long Lythe and the Short Lythe, that took him through the beechwoods and beside the stream. During the years that he paced along them, James Bruce, the Scot, travelled to the source of the Blue Nile, and Captain Cook sailed across the Pacific. White knew all about these exciting discoveries. He had met and corresponded with Sir Joseph Banks, the great naturalist who was with Cook on his first voyage. But Selborne and its countryside gave White all the wonder, splendour and drama that

FIG.5 (below)
Eric Ravilious, *Memorial to Gilbert White*, 1938. Headpiece of volume one and tailpiece of volume two. Wood-engraving on paper. Illustration to *The Writings of Gilbert White of Selborne,* H.J. Massingham, ed., London: Nonesuch Press, 1938.

FIG.6 (opposite)
Frederick Marns, *Gilbert White's Path, from the Back Door of 'The Wakes'*. Illustration to Gilbert White, *The Natural History of Selborne*, with preface by Lord Selborne and an introduction by Sir David Attenborough, London: Shepheard-Walwyn, 1977.

he required. The deep lanes that led from the village he found very grotesque and wild. 'These rugged gloomy scenes affright the ladies when they peep down into them from the paths above and make timid horsemen shudder when they ride along them,' he wrote, adding, typically, 'but they delight the naturalist with their various botany.'

A mile from his home lay Wolmer Forest, 'a hungry, sandy, barren waste' with a bog and a pond which White regarded as 'a vast lake'. It was less than a mile across. Through the forest browsed red deer. Queen Anne was said to have seen a herd of five hundred and that, noted White, was 'a sight worthy of the attention of the greatest sovereign'. The South Downs, through which he used to ride on annual visits to his aunt at Ringmer, he called 'majestic mountains,' and doubtless they were as awe-inspiring to him as the Himalayas might be to our more jaded eyes.

Gilbert White made no boastful bids for immortality. It took him a life-time to get round to publishing his single book. But just here and there in the village that meant so much to him, he put an inscription – his two initials. He then had them carved on the melon wall that he built in his garden, on the extension that he added to The Wakes, on his brewhouse and on the desk in his study.

And that, too, was all that he asked for on his tombstone – an epitaph that was modest, unobtrusive and truthful, like his enduring book.

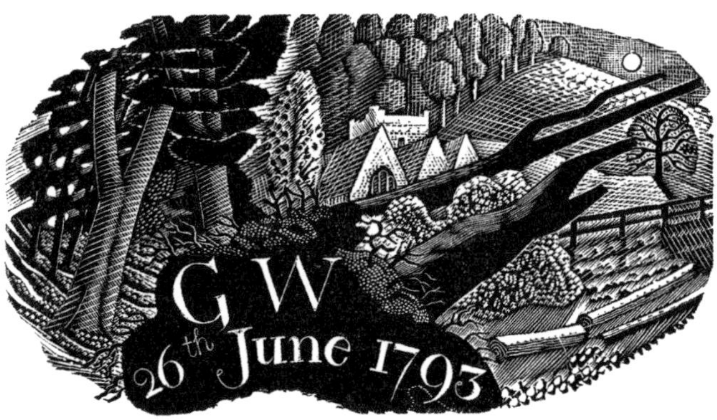

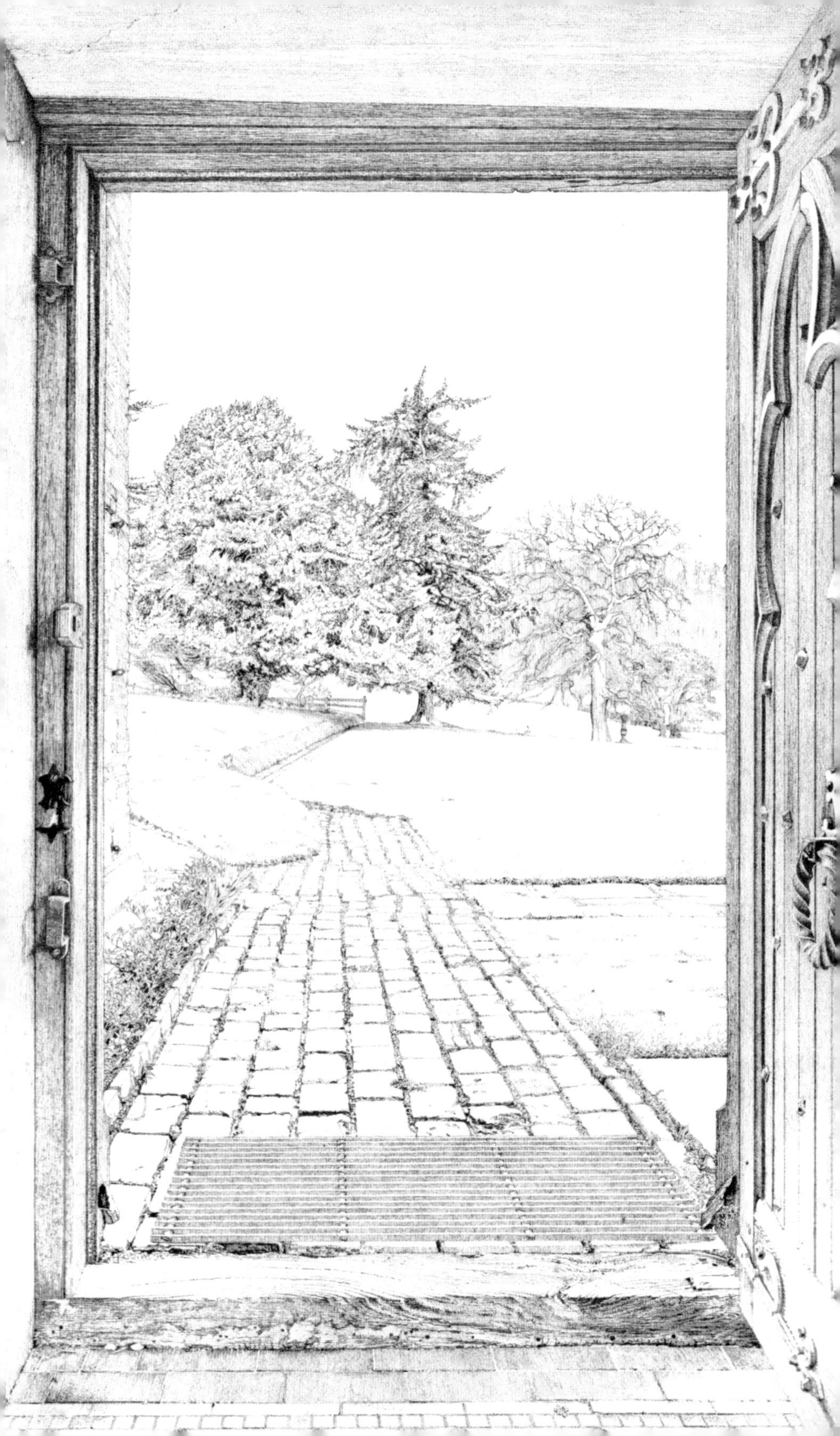

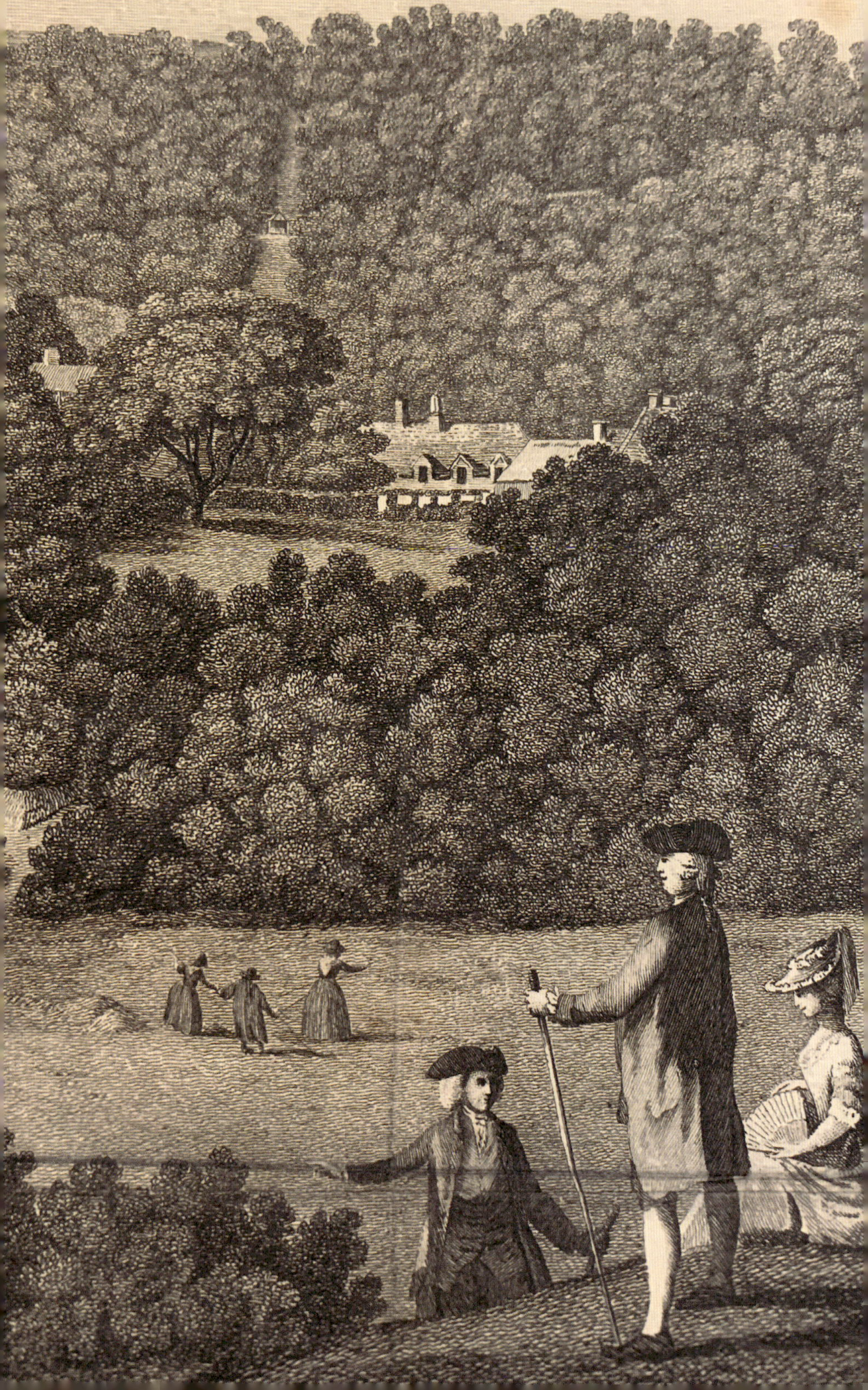

'AN INVITATION TO SELBORNE'

Gilbert White, 1745

See, Selborne spreads her boldest beauties round
The varied valley, and the mountain ground,
Wildly majestic! What is all the pride
Of flats, with loads of ornament supply'd?
Unpleasing, tasteless, impotent expense,
Compared with nature's rude magnificence.

Arise, my stranger, to these wild scenes haste,
The unfinish'd farm awaits your forming taste:
Plan the pavilion, airy, light and true;
Thro' the high arch call in the length'ning view;
Expand the forest sloping up the hill;
Swell to a lake the scant, penurious rill;
Extend the vista, raise the castle mound
In antique taste, with turrets ivy-crown'd;
O'er the gay lawn the flow'ry shrub dispread,
Or with the blending garden mix the mead;
Bid China's pale, fantastic fence, delight,
Or with the mimic statue trap the sight.
Oft on some evening, sunny, soft and still,
The muse shall lead thee to the beech-grown hill,
To spend in tea the cool, refreshing hour,
Where nods in air the pensile, nest-like bower;
Or where the Hermit hangs the straw-clad cell,

FIG.7
Detail from *North East View of Selborne* (see fig.14).

Emerging gently from the leafy dell;
By fancy plann'd; as once th' inventive maid
Met the hoar sage amid the secret shade;
Romantic spot! from whence in prospect lies
Whate'er of landscape charms our feasting eyes;
The pointed spire, the hall, the pasture-plain,
The russet fallow, or the golden grain,
The breezy lake that sheds a gleaming light,
Till all the fading picture fail the sight.

Each to his task; all different ways retire,
Cull the dry stick; call forth the seeds of fire;
Deep fix the kettle's props, a forky row,
Or give with fanning bat the breeze to blow.

Whence is this taste, the furnish'd hall forgot,
To feast in gardens, or th'unhandy grot?
Or novelty with some new charms surprises,
Or from our very shifts some joy arises.
Hark, while below the village-bells ring round,
Echo, sweet nymph, returns the soften'd sound;
But if gusts rise, the rushing forests roar,
Like the tide tumbling on the pebbly shore.

Adown the vale, in lone, sequester'd nook,
Where skirting woods imbrown the dimpling brook,
The ruin'd Convent lies; here wont to dwell
The lazy canon midst his cloister'd cell;
While papal darkness brooded o'er the land,
Ere reformation made her glorious stand:
Still oft at eve belated shepherd-swains
See the cowl'd spectre skim the folded plains.
To the high temple would my stranger go,
The mountain-brow commands the woods below;
In Jewry first this order found a name,

When madding Croisades set the world in flame;
When western climes, urg'd on by pope and priest,
Pour'd forth their millions o'er the deluged east;
Luxurious knights, ill suited to defy
To mortal fight Turcéstan chivalry.

Nor be the Parsonage by the muse forgot.
The partial bard admires his native spot;
Smit with its beauties, loved, as yet a child,
(Unconscious why) its scapes grotesque, and wild.
High on a mound th' exalted gardens stand,
Beneath, deep vallies scoop'd by Nature's hand.
A Cobham here, exulting in his art,
Might blend the General's with the Gardener's part;
Might fortify with all the martial trade
Of rampart, bastion, fosse, and palisade;
Might plant the mortar with wide threatening bore,
Or bid the mimic cannon seem to roar.

Now climb the steep, drop now your eye below,
Where round the blooming village orchards grow;
There, like a picture, lies my lowly seat,
A rural, shelter'd, unobserved retreat.

Me far above the rest Selbornian scenes,
The pendent forests, and the mountain-greens
Strike with delight; there spreads the distant view,
That gradual fades till sunk in misty blue:
Here nature hangs her slopy woods to sight,
Rills purl between and dart a quivering light.

'THE NATURALIST'S SUMMER-EVENING WALK'

Gilbert White, 1769

' - *equidem credo, quia sit divinitùs illis*'
('- *I deem their breasts inspired with a divine sagacity*')
VIRGIL, Georgics, i.415

When day declining sheds a milder gleam,
What time the May-fly haunts the pool or stream;
When the still owl skims round the grassy mead,
What time the timorous hare limps forth to feed;
Then be the time to steal adown the vale,
And listen to the vagrant cuckoo's tale;
To hear the clamorous curlew call his mate,
Or the soft quail his tender pain relate;
To see the swallow sweep the dark'ning plain
Belated, to support her infant train;
To mark the swift in rapid giddy ring
Dash round the steeple, unsubdu'd of wing:
Amusive birds! – say where your hid retreat
When the frost rages and the tempests beat;
Whence your return, by such nice instinct led,
When spring, soft season, lifts her bloomy head?
Such baffled searches mock man's prying pride,
The God of Nature is your secret guide!

While deep'ning shades obscure the face of day
To yonder bench leaf-shelter'd let us stray,
'Till blended objects fail the swimming sight,
And all the fading landscape sinks in night;
To hear the drowsy dorr come brushing by
With buzzing wing, or the shrill cricket cry;
To see the feeding bat glance through the wood;
To catch the distant falling of the flood;
While o'er the cliff th'awakened churn-owl hung
Through the still gloom protracts his chattering song;
While high in air, and pois'd upon his wings,
Unseen, the soft, enamour'd woodlark sings:
These, Nature's works, the curious mind employ,
Inspire a soothing melancholy joy:
As fancy warms, a pleasing kind of pain
Steals o'er the cheek, and thrills the creeping vein!

Each rural sight, each sound, each smell, combine;
The tinkling sheep-bell, or the breath of kine;
The new-mown hay that scents the swelling breeze,
Or cottage chimney smoking through the trees.

The chilling night-dews fall: – away, retire;
For see, the glow-worm lights her amorous fire!
Thus, ere night's veil had half obscur'd the sky,
Th'impatient damsel hung her lamp on high:
True to the signal, by love's meteor led,
Leander hasten'd to his Hero's bed.

'A HARVEST SCENE'

Gilbert White, date unknown

Wak'd by the gentle gleamings of the morn,
Soon clad, the reaper, provident of want
Hies cheerful hearted to the ripen'd field;
Nor hastes alone; attendant by his side
His faithful wife, sole partner of his cares,
Bears on her breast the sleeping babe; behind,
With steps unequal, trips her infant train:
Thrice happy pair, in love and labour join'd! –

All day they ply their task; with mutual chat
Beguiling each the sultry, tedious hours:
Around them falls in rows the sever'd corn,
Or the shocks rise in regular array.

But when high noon invites to short repast,
Beneath the shade of sheltering thorn they sit,
Divide the simple meal, and drain the cask:
The swinging cradle lulls the whimpering babe
Meantime; while growling round, if at the tread
Of hasty passenger alarm'd, as of their store
Protective, stalks the cur with bristling back,
To guard the scanty scrip and russet frock.

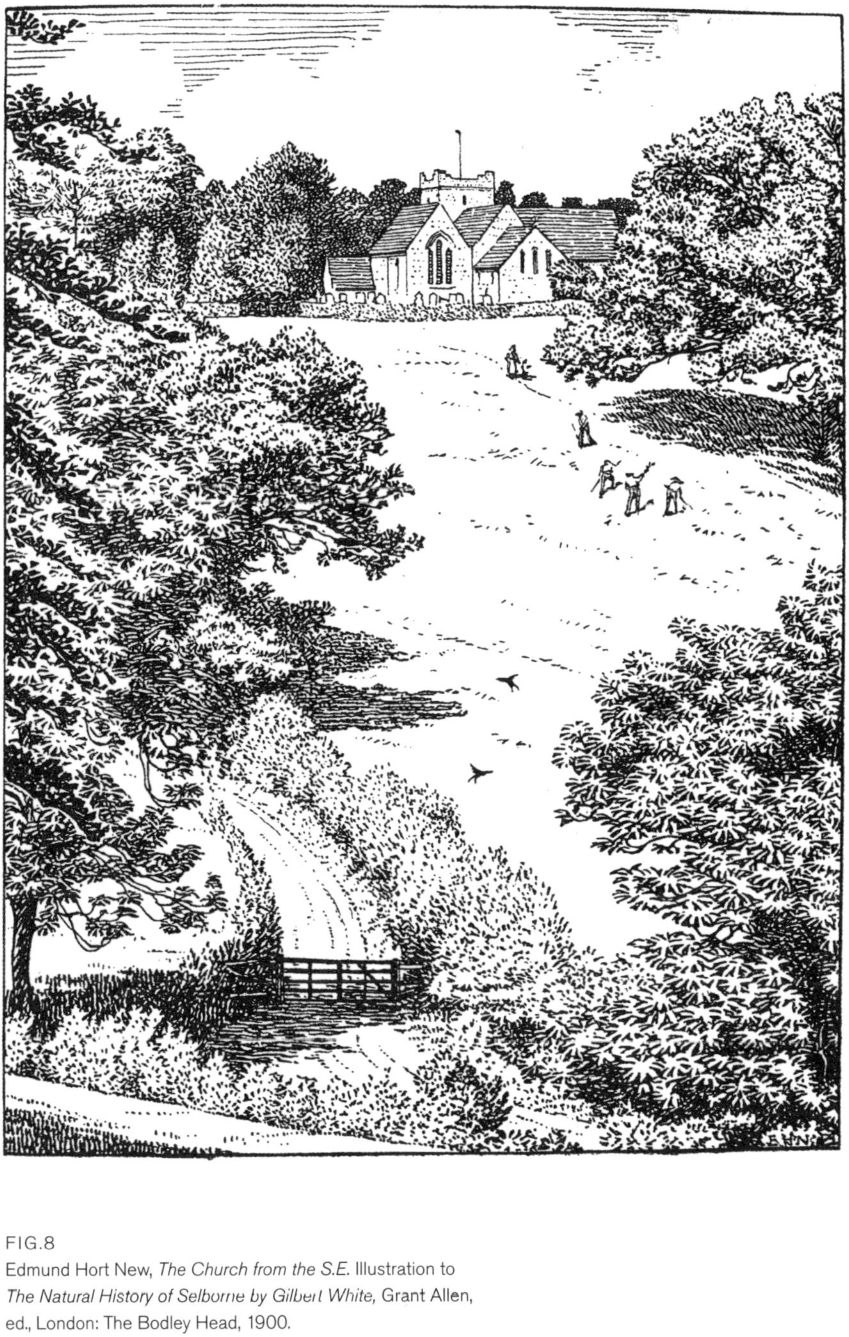

FIG.8
Edmund Hort New, *The Church from the S.E.* Illustration to *The Natural History of Selborne by Gilbert White,* Grant Allen, ed., London: The Bodley Head, 1900.

The writings of GILBERT WHITE of Selborne

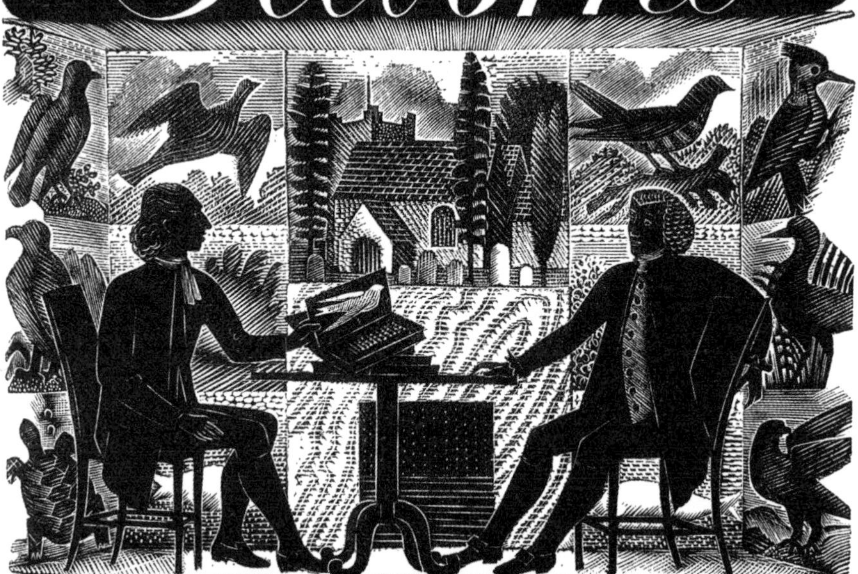

Selected and edited with an introduction by

H. J. Massingham

VOLUME I. LONDON 1938

The Nonesuch Press

2.
OBSERVING NATURE: THE ILLUSTRATORS OF GILBERT WHITE'S NATURAL HISTORY OF SELBORNE

Simon Martin

If people that live in the country would take a little pains,
daily observations might be made with respect to animals,
and particularly regarding their actions and economy,
which are the life and soul of natural history.
Gilbert White, 12 May 1770

Sitting quietly behind some of our greatest cultural achievements is the observation of nature. More than simply noticing something, observation requires us to use skills of perception and to recognise the significance of the subject under scrutiny. Although it is something we do instinctively, observation requires presence of mind, a sensitivity to subtle changes and to seemingly insignificant details. It has been essential to the collecting and recording of data which led to some of the most important scientific discoveries, from gravity to the theory of evolution. Countless stanzas of poetry and expressive passages of paint would not have come into being without the attention given to nature, even for the shortest periods of time. With John Constable we are conscious of the oak trees, the streams and the passing clouds over

FIG.9
Eric Ravilious, Title page of volume one, 1937. Wood-engraving on paper. Illustration to H.J. Massingham, ed., *The Writings of Gilbert White – Selected and Edited with an Introduction by H.J. Massingham*, London: The Nonesuch Press, 1938.

Dedham Vale; with William Wordsworth we think of dancing daffodils and the lakes; with Isaac Newton we think of a falling apple; and with the Brontë sisters we are conscious of the ravages of the weather on the moors.

Whilst the relationship of man to nature has been an enduring theme in English art and literature, nature *itself* has largely been the supporting act, or rather the backdrop against which the narrative is played out. In the writings of the Rev. Gilbert White (1720–93), however, the natural world forms not only the setting for the action, but also the subject. In his *Natural History of Selborne*, first published in 1789, White presented a way of looking at nature that was reverent and attentive, expressing his innate curiosity about the birds and animals in a single parish over a period of 40 years. The non-human inhabitants of the village of Selborne hold the focus of our attention, whilst the humans are largely present as observers or incidental characters within the narrative. Selborne is home to one of the most particular, and sometimes fleeting, cast of characters in English literature – including a pair of hoopoe birds, a hungry toad, a barn owl, a pony called Mouse and a rather fractious tortoise named Timothy. Yet, these are not fictional animals brought to life in the manner of children's stories, such as the characters in *The Wind in the Willows, The Tales of Peter Rabbit* or *Tarka the Otter*. Instead, the creatures of Selborne were real beings observed from life, recorded in minute detail by an equally unlikely literary hero – an eighteenth-century parson–naturalist.

What is remarkable is that not only do the avian, animal and reptilian characters occupy a place within the narrative structure of an epistolary novel, but White's writing is a rare combination of scientific precision, poetic narrative and conversational anecdote. So often he is writing about seemingly unimportant and unprepossessing topics, the 'minute particulars' which take us deep into the hedgerows and hollow lanes of his village. It is no wonder that it has had such an appeal for artists over the subsequent two centuries; rarely have birds and animals in literature been so well described, and with such character, providing a perfect brief for an illustrator to take as their starting point.

With simplicity and precision, White recorded the natural life on his doorstep in Hampshire, with occasional forays into Sussex, 'watching narrowly', as he was to describe it. Unlike his grandfather (also the Rev. Gilbert White), he could never become the vicar of Selborne for he was a Fellow of Oriel College Oxford, and the 'living' for the parish was with Magdalen College. But at four different times he was curate of Selborne, as well as of other parishes, such as Farrington just two miles away. At his home in the village, known as The Wakes, he recorded his experiments in landscape design and horticulture in his *Garden Kalendar*, but also, more importantly, his observations of the area's ecology in a series of letters to his fellow naturalists Thomas Pennant (1726–98) and the Hon. Daines Barrington (1727–1800).

White first wrote to Pennant in 1767, and in the following year began keeping his records in his *Naturalist's Journal*. Designed by Barrington, it was printed by Gilbert's brother Benjamin White, who ran the publishing firm Horace's Head in Fleet Street (which also printed Pennant's books). Benjamin introduced his brother to Barrington and they had started exchanging letters in 1769. White was an enthusiastic amateur, whereas his correspondents were Fellows of the Royal Zoological Society. Yet, somehow, the immediacy of his voice and his simple joy and curiosity about nature has meant that we remember and cherish his writing today.

White compiled, ordered and composed his carefully transcribed correspondence with Pennant and Barrington into his book *The Natural History and Antiquities of Selborne, in the County of Southampton with Engravings, and an Appendix.* It was printed by Thomas Bensley (1759–1835) and published by his brother Benjamin.[1] Additional 'letters' were written in order to tie the book together, with a parish history known as the 'Antiquities of Selborne' spanning the Bronze Age, the Romans and beyond, reflecting the growing interest in antiquarian study in the eighteenth century. Perhaps surprisingly, given its geographically narrow view of a small English village, the book was an immediate success and has never since been out of print. With over 400 editions and reissues, it is apocryphally said to be the fourth most-published book in English, after the Bible, the works of Shakespeare, and John Bunyan's *The Pilgrim's Progress*. Whether this is strictly true hardly matters, for although the life and writings of Gilbert White may seem parochial (in the truest sense of the word), their impact has been expansive and profound, with his reach felt around the globe 300 years after his birth.

Writing to Barrington, White declared that he was an 'outdoor naturalist, one that takes his observations from the subject itself, and not from the writings of others'.[2] Whereas his contemporaries

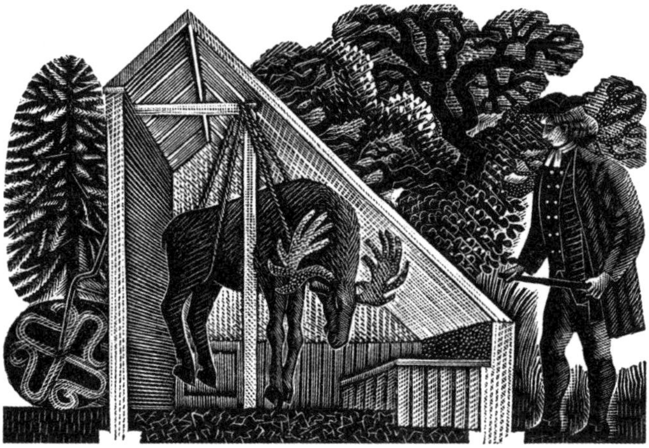

FIG.10
Eric Ravilious, *Duke of Richmond's Greenhouse*, 1938. Wood-engraving on paper. Illustration to H.J. Massingham, ed., *The Writings of Gilbert White – Selected and Edited with an Introduction by H.J. Massingham*, London: The Nonesuch Press, 1938.

would stay indoors, conducting detailed examinations of dead specimens in their studies, White believed in *observation* rather than *collection* – watching living birds and animals in their natural habitats, using Linnean classifications. In March 1770, he wrote of being 'greatly disappointed' on arriving at Goodwood to see a female moose belonging to the Duke of Richmond, which had been brought over from North America, only to find that it had died on the previous morning. Nevertheless, White had proceeded to examine and measure the beast, which had been slung up in an old greenhouse, in order to understand its distinction from the species of deer that he had previously encountered in the wild (fig.10). However, he bemoaned that, 'though it had been dead so short a time, it was in so putrid a state that the stench was hardly supportable.'[3] This idiosyncratic episode is a comic illustration of White's earnest and assiduous dedication. He disliked animals in captivity, declaring that he was 'no bird catcher; and so little used to birds in cages.'[4] For these reasons and others, White has been described as 'the first ecologist'. His writings made a deep impression on Charles Darwin as a young man, who read the book in 1837 and later made a pilgrimage to Selborne.[5] It inspired Darwin to take 'much pleasure in watching the habits of birds' and to wonder, 'why every gentleman did not become an ornithologist'.[6] Other notable admirers include the naturalists Gerald Durrell and Sir David Attenborough, who has described White as a man 'in total harmony with his world'.[7]

It was Gilbert White who made the first field observations to prove the existence of three kinds of leaf-warblers – the chiff-chaff, the willow-warbler and the wood-warbler – largely on the basis of their different songs. He also discovered and named the lesser white-throat ('A rare and I think new little bird frequents my garden... This bird much resembles the white-throat, but has a more white or rather silvery breast and belly; is restless and active, like the willow wrens, and hops from bough to bough examining every part for food');[8] the harvest mouse ('they hang their nests for breeding up amidst the straws of the standing corn... yet I find that in the winter they burrow deep in the earth and make warm beds of grass');[9] and the noctule bat ('that large species of bat which I call *Verspertilio altivolans* from its manner of feeding high in the air').[10] Such was his focus that instead of discussing the tumultuous political events of his times, such as the Jacobite Rebellion or the new-found independence of the United States, White writes about nightjars and the copulation of frogs. He does, however, briefly acknowledge the French Revolution, which broke out in the year his book was first published. He recorded on 21 January 1793: 'Thrush sings, the song-thrush: the missel-thrush has not been heard. On this day Louis XVI, late king of France, was beheaded at Paris, & his body flung into a deep grave without any coffin, or funeral service performed.'[11]

Like others before us, we are conscious that the world White describes is disappearing or has already gone. Writing in 1874, almost a century after the book was published, in his introduction to White's *Natural History*, James Harting observed:

> No less remarkable are the changes which have taken place in the flora and fauna of the district. The red-deer, which once roamed the Forest of Wolmer, and which were driven 'along the vale' in a herd of 500 for the amusement of Queen Anne have long since become extinct... Those noble birds the bustards, which once frequented the downs, and which seen in flocks at a distance were thought to resemble fallow-deer, have entirely vanished as denizens of England.[12]

Fast forward then to the 2020s, when a climate emergency has been declared, with international protests by Extinction Rebellion. With forests burning in the Amazon and bushfires in Australia, not to mention a global pandemic, it is saddening to realise how much else has been lost in the two centuries since White described his natural world. Although his letters were written in another age, with far different politics and ideals from today, White's observations still have resonance in the present moment. He recognised that all creatures had a role to play, no matter how great or small. As he wrote in May 1770, 'Earthworms, though in appearance a small and despicable link in the chain of nature, yet, if lost, would make a lamentable chasm. For, to say nothing of half the birds, and some quadrupeds, which are almost entirely supported by them, worms seem to be the great promoters of vegetation, which would proceed but lamely without them'.[13] White reminds us that in order to care for and preserve nature, we must first notice it.

Gilbert White mixed what his biographer Richard Mabey described as 'scientific jargon and rural commonplace' in a readable literary style.[14] There is such an immediacy to White's writing that he simultaneously seems to be *of* his time and yet timeless – a presence at our sides whilst we are reading. 'He escapes from Selborne, from his own age,' declared Virginia Woolf in 1939, 'winging his way to us in the dusk along the hedgerows'.[15] Always having an interest in both the form of letter-writing and in the novel, Woolf focussed on what she saw as the novelistic qualities of White, declaring how, 'By some apparently unconscious device of the author has a door [been] left open, through which we hear distant sounds, a dog barking, cart-wheels creaking…'[16] That vitality comes in part from White's verbal precision; because he was seeking the most appropriate analogies, his prose is full of vivid observations.

White wrote poetry, some of which was included in the *Natural History*, but his formal verse was not the reason why the book has had a special appeal for poets, such as Henry David Thoreau or Samuel Taylor Coleridge, who called it 'a sweet, delightful book'.[17] Instead, it is the poetic quality of his descriptions. He observes 'vast, swaggering, rock-like clouds'[18] and the 'shapely figured aspect of the chalk hills.'[19] He describes a goldcrest as 'the golden-crowned wren whose crown glitters like burnished gold',[20] whilst paths and fields after heavy rain are 'all in a batter'.[21] Writing of rooks at their 'evening proceedings,' he recalls how they made 'rather a pleasing murmur, very engaging to the imagination, and not unlike the cry of a pack of hounds in hollow echoing woods, or the rushing of the wind in tall trees, or the tumbling of the tide upon a pebbly shore'.[22]

As Richard Mabey observes:

> What is striking is the way Gilbert often arranges his sentence structure to echo the physical style of a bird's flight. So 'The white-throat uses odd jerks and gesticulations over the tops of hedges and bushes'; and 'Woodpeckers fly *volatu undosu* [in an undulating flight], opening and closing their wings at every stroke, and so are always rising and falling in curves.[23]

In 1935, an edition of the book was presented 'rather as a literary masterpiece than as a scientific treatise,' and celebrated White's 'miniature poetry'.[24] During the hot summer of 1973, shortly before his death, W.H. Auden penned the poem 'Posthumous Letter to Gilbert White' (see p.92), in which he confessed how, 'Selfishly, I, too, would have plumbed to know you. I could have learned so much'.[25]

Recognising White's unique ability to convey the character of the creatures he describes, in 1946 Sylvia Townsend Warner wrote *The Portrait of a Tortoise,* a compilation of White's accounts of his pet Timothy whom he inherited from his aunt Rebecca Snooke at Ringmer (figs 11, 12); Warner observes that 'Gilbert White's private love-affair with Nature runs like a chalk-stream river, pellucid, tranquil, and irresistible'.[26] White is even referenced in magical fiction. In T.H. White's mystical Arthurian romance

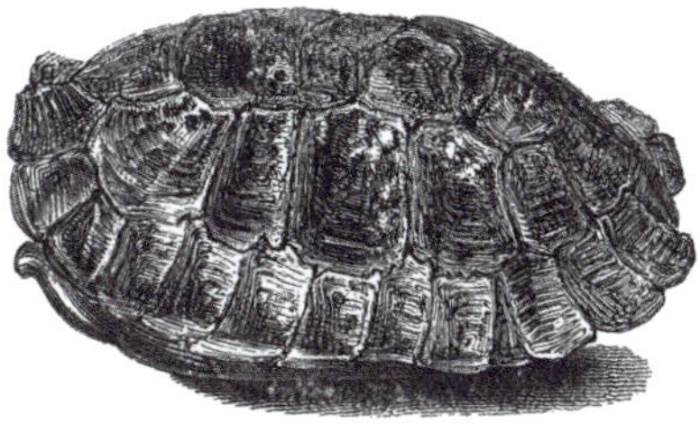

THE SHELL OF GILBERT WHITE'S TORTOISE.

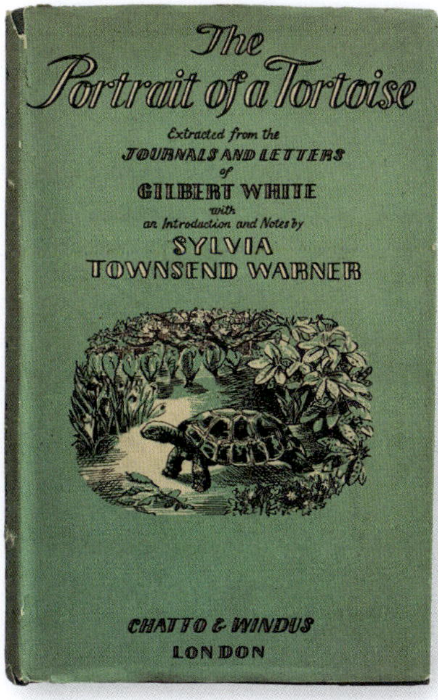

FIG.11
Unknown artist, *The Shell of Gilbert White's Tortoise*. Wood-engraving on paper, from *The History of British Birds* reproduced in the James Harting edition of *The Natural History of Selborne*, London: Bickers & Son, 1875.

FIG.12
Irene Hawkins (1906–1979), cover of Sylvia Townsend Warner, ed. *The Portrait of a Tortoise: Extracted from the Journals and Letters of Gilbert White*, London: Chatto & Windus, 1946.

The Once and Future King, published in 1958, the young Wart asks Merlyn and Archimedes the owl whether birdsong is language. Looking forward to the eighteenth century from the perspective of the Middle Ages, the wizard predicts how Gilbert White, 'remarks, or will remark, however you like to put it, that "the language of birds is very ancient, and, like other ancient modes of speech little is said, but much is intended".'[27] It is an instance of how Gilbert White's evocative words have not just inspired naturalists and scientists, but seeped into popular consciousness, into language and literature.

It seems clear why White has appealed so strongly to artists and poets, with his sensitive, richly descriptive prose and his evocation of the *genius loci* (spirit of place) – an idea which has long preoccupied writers and artists, and with which White would have been familiar from Alexander Pope's writings on garden and landscape design. In the seventeenth and eighteenth centuries, local histories and chorography (the describing or mapping of a region or district) had become increasingly popular. Written by the English antiquarian, historian and topographer William Camden (1551–1623), the first chorographic survey of the islands of Great Britain and Ireland, *Britannia,* was published in 1586. Subsequent notable antiquaries included Sir William Dugdale (1605–85), who wrote *Antiquaries of Warwickshire,* and John Aubrey (1626–97), who recorded the megalithic and other field monuments in southern England in his *Monumenta Britannica* (written between 1663 and 1693) and wrote *Hypomnemata Antiquaria,* a county history of Warwickshire. This was modelled on the *Natural History of Oxfordshire* (published 1677) by Robert Plot, the first keeper of the Ashmolean Museum. The growing interest in antiquarian history had been reflected in the reformation of the Society of Antiquaries, whose first minuted meeting took place in 1718 and who published illustrated papers known as the *Vetusta Monumenta.*[28] In 1786, a facsimile of the Doomsday Book was published, indicative of the interest in compiling collections at parish, urban and county level.

This gentlemanly 'topophilia' (love of particular places) led to the writing of parish histories, many of which were never published but provided an occupation for gentlemen and for clergy who often had a vested interest in the subject matter. The clergy had the education, social position and leisure to undertake such pursuits, with a notable example being the Rev. Philip Morant (1700–70) who compiled histories of Colchester and Essex. The Rev. Gilbert White, however, was to take an established form of parish history and chorography and transform it through an epistolatory structure, viewing antiquities and natural history through a personal lens.

The landscape painter John Constable R.A. (1776–1837) wrote admiringly of White's *Natural History* in 1821: 'It only shows what a real love for nature will do – surely the serene and blameless life of Mr White, so different from the folly and quackery of the world, must have fitted him for such a clear and intimate view of nature.'[29] It is difficult to ascertain White's own views on art, although he was capable of writing in a self-consciously painterly style, as he did on 26 October 1783 when declaring:

> If a masterly land-scape painter was to take our hanging woods in their autumnal colours, persons unacquainted with the country would object to the strength and deepness of the tints and would pronounce at an exhibition that they were heightened and shaded beyond Nature. Wonderfully lovely to the Imagination are the colourings of our woodland scapes at this season of the year.[30]

With the modest income of a parson, White was not necessarily in the position to collect landscape paintings, nor to commission portraits of himself from the leading artists of the day, as did his brothers or Thomas Pennant, whose portrait was painted in 1776 by Thomas Gainsborough R.A. (1727–88), or Daines Barrington who was painted by both Samuel Drummond A.R.A. (1766–1844) and Joseph Slater Snr. (*c.*1750–1805). The posthumous portraits of White by artists have therefore been re-imaginings of how he might have looked.

The illustrations for his book, however, did require White to take on the role of commissioner. Before the publication of the *Natural History*, White was persuaded to look for an artist to illustrate his book by his publisher brother Benjamin, who was conscious of the increasing interest in illustration and the corresponding potential sales benefit. Gilbert had himself encouraged his other brother, John, a fellow naturalist based in Gibraltar, to include as many plates as possible in the book he was planning, the now-lost *Fauna Calpensis*. Picture-books had become highly fashionable by this time. Benjamin White had published White's correspondent Thomas Pennant's *British Zoology* in folio between 1761 and 1766 with 132 plates of birds

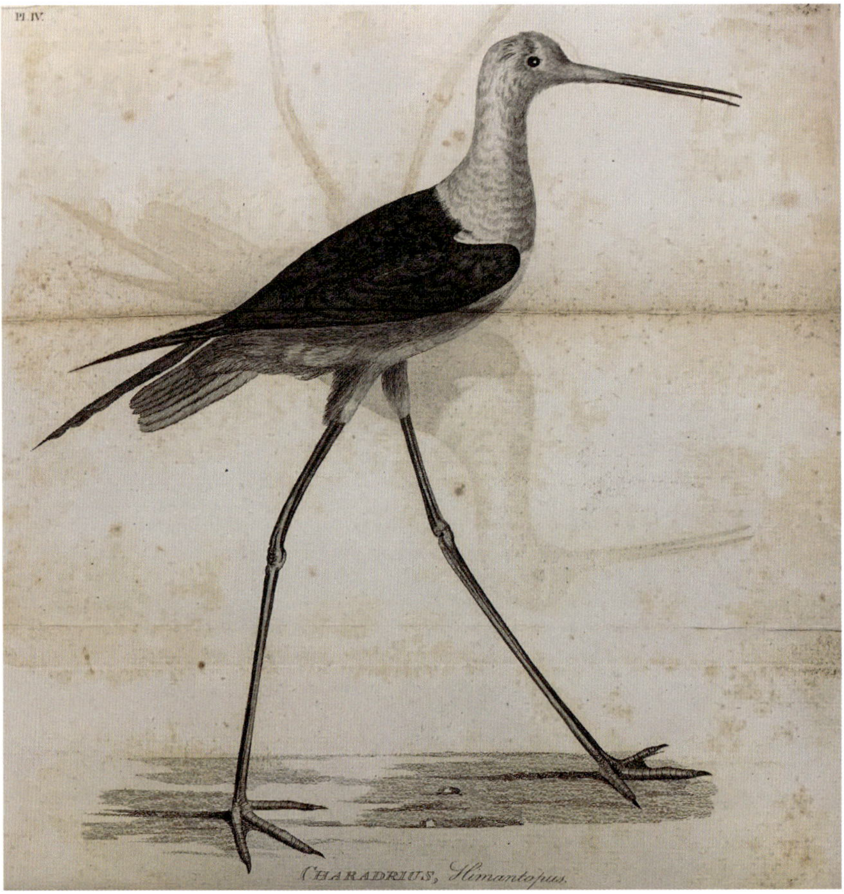

FIG.13
Artist unrecorded (possibly Peter Paillou or Moses Griffith), *Black-winged stilt – Charadrius himantopus.* Engraving on paper, printed by Thomas Bensley. Illustration to the first edition of *The Natural History and Antiquities of Selborne, in the County of Southampton with Engravings, and an Appendix,* London: B. White & Co., 1789.

and quadrupeds (four-legged animals), later expanding it into four volumes with reptiles, fish and molluscs. The full-plate images in Pennant's book, printed by W. Eyres, have a charm derived from their naïve simplification. During the eighteenth century there was a surge of interest in natural history illustration, with the cataloguing and visual recording of nature receiving impetus from the likes of the Swedish botanist and zoologist Carl Linneus (1707–78) who was in correspondence with White's brother John. Others included George Clifford III (1685–1760), a wealthy Anglo-Dutch merchant, and Sir Hans Sloane (1660–1753), the founder of the British Museum, who commissioned artists such as George Edwards (1694–1773) and Georg Dionysus Ehret (1708–70) as natural history painters. Edwards was known as 'the father of British ornithology' and had published the first volume of his illustrated *A Natural History of Uncommon Birds* in 1743, with subsequent volumes under the title *Gleanings of Natural History* issued in the years before 1764, including engravings and descriptions of over 600 subjects in natural history that had not previously been delineated. This was also a period of geographic exploration during which Sir Joseph Banks (1743–1820) and Daniel Solander (1733–82) accompanied Captain Cook's first voyage to the Pacific on the *Endeavour* in 1768. Hired by Banks, Sydney Parkinson (1745–71) completed 674 detailed drawings and 269 watercolours on this journey, to illustrate Banks' *Florilegium* (published many years later).

Given the focus of White's own book it is perhaps surprising that, rather than commissioning natural history illustration, he settled on a largely topographical approach. In the first edition of *The Natural History of Selborne* there is only one image that could be conceived as constituting natural history illustration: a stilt plover (*Charadrius himantopus*) printed by Thomas Bensley (1760–1835) (fig.13). It seems possible that this was based on a painting by either Peter Paillou (*c.*1720–90) or Moses Griffith (1749–1819), who were both employed by Pennant to paint pictures of birds, many of which were engraved by Peter Mazell (fl. 1761–97), who also engraved images in *The Natural History of Selborne*.[31] Quite why the stilt plover was particularly selected from the many birds mentioned in the book is not entirely clear, but it was presumably singled out due to the peculiar appearance of its unfeasibly long legs. As White was to write:

> It must be matter of great curiosity to see the stilt plover move; to observe how it can wield such a length of lever with such feeble muscles as the thighs seem to be furnished with. At best one should expect it to be but a bad walker: but what adds to the wonder is that it has no back toe. Now without that steady prop to support its steps it must be liable, in speculation, to perpetual vacillations, and seldom able to preserve the true centre of gravity.[32]

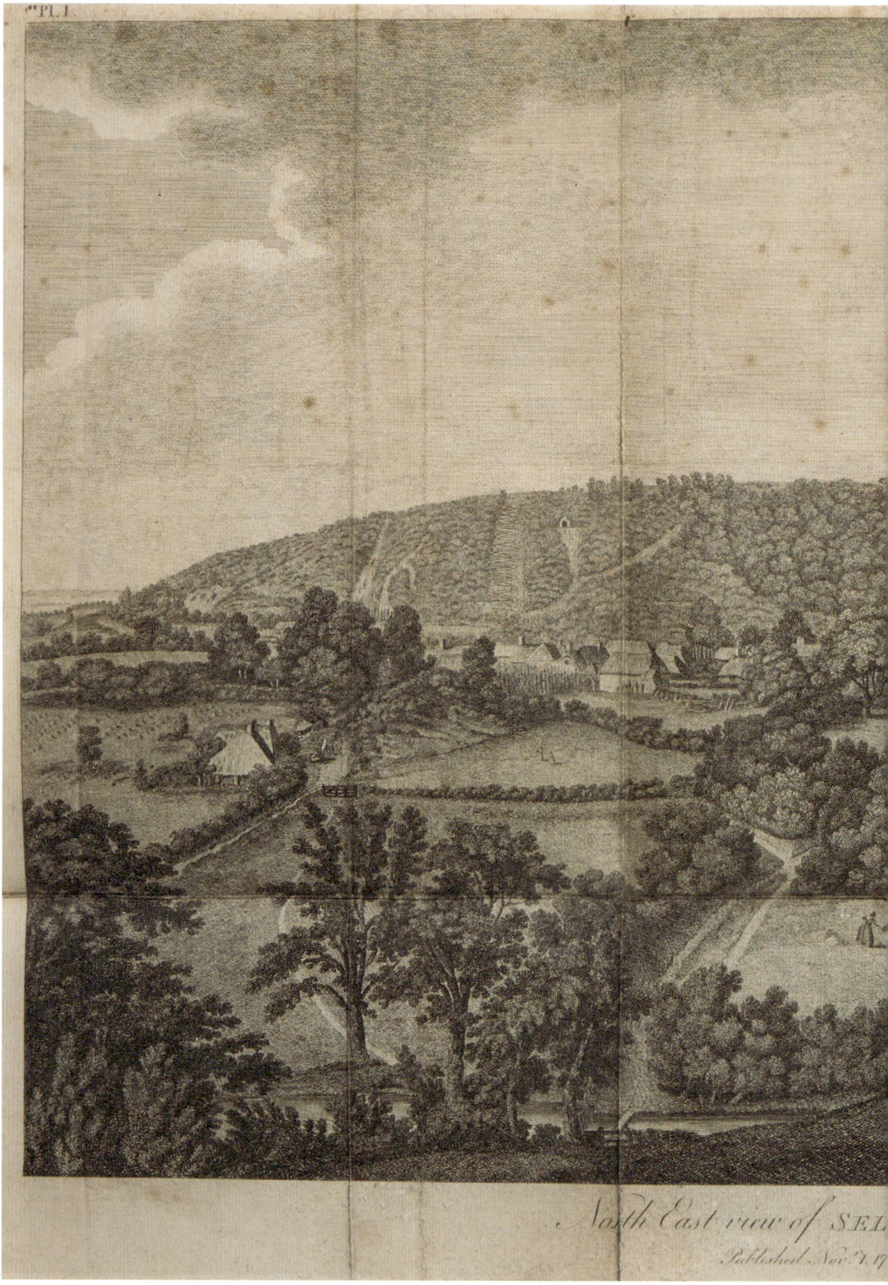

FIG.14

Samuel Hieronymus Grimm, *North East View of Selborne, from the Short Lythe.*
Etching on paper. Illustration to the first edition of Gilbert White, *The Natural History and Antiquities of Selborne, in the County of Southampton with Engravings, and an Appendix*, London: B. White & Co., 1789.

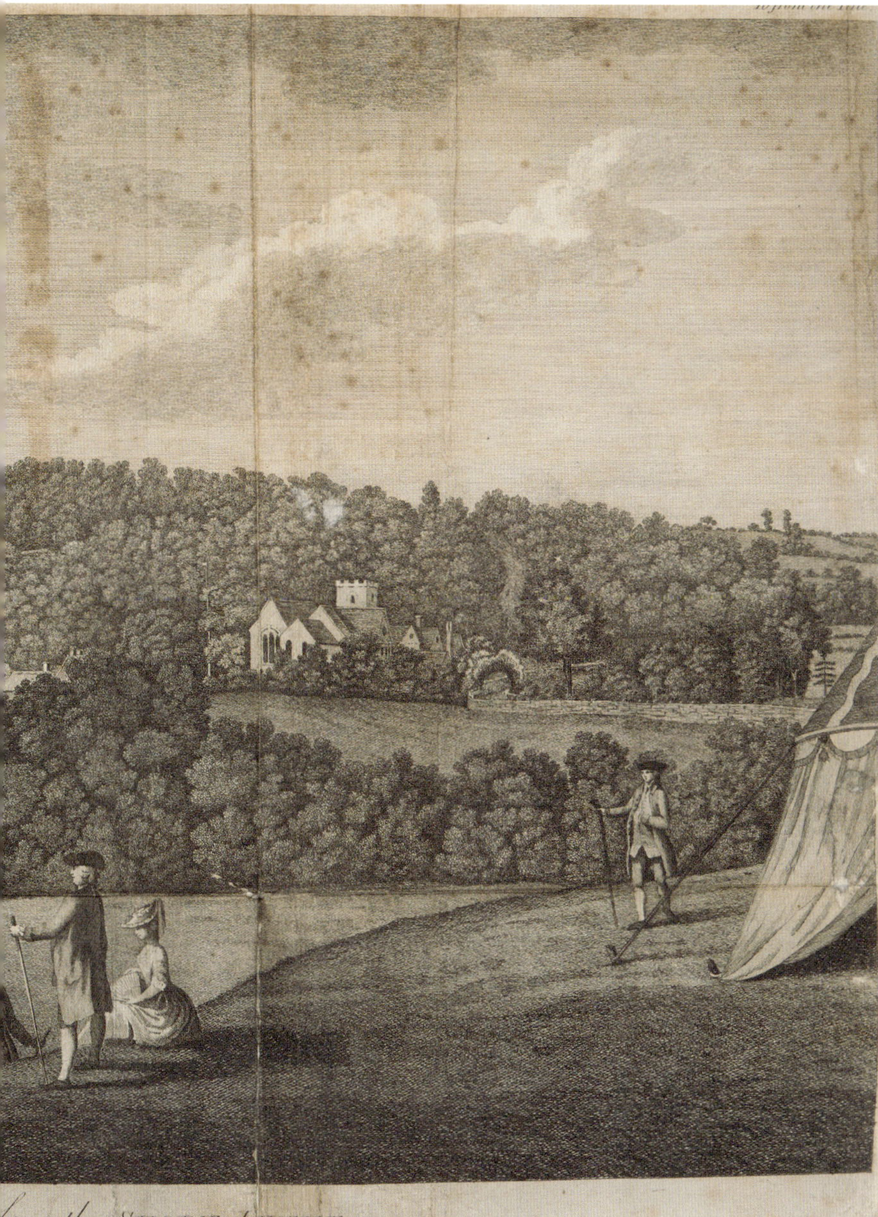

from the SHORT LYTHE.

reets, by B.White & Son.

Instead of conventional natural history illustration White commissioned a set of topographical landscape drawings from the landscapist Samuel Hieronymus Grimm (1733–94), a Swiss artist who had come to specialise in showing England to the English. Since his arrival in the country in 1768 Grimm had worked for several notable patrons, including the lawyer and antiquary Sir William Burrell, for whom he made over 900 drawings of Sussex between 1780 and 1791, and the Rev. Sir Richard Kaye, who set about recording the history and antiquities of England and tasked him with recording 'everything curious'. Together with Edward Edwards, Grimm had been commissioned to produce watercolours for the Society of Antiquaries' *Vetusta Monumenta* in the 1770s. Notwithstanding Grimm's other patrons, White described him as 'my artist' and praised his 'genius, assiduity and modest bearing'.[33] The parson's friend Thomas Mulso expressed concern 'that Ornament, which You seemed to set your heart upon'[34] would delay the book, and he was to joke that White was behaving like 'an Italian Magnifico, with your Designer at your Elbow.'[35] Although it is perhaps curious that White engaged a topographical artist, rather than a painter of animals or a scientific illustrator, he approved of aspects of Grimm's work: 'His buildings, human figures, quadrupeds, waters, perspective among trees, are good; but his trees are not so pleasing: he also has a vein of humour, but I shall not allow him to call it forth, as all my plates must be serious.'[36]

Arriving in Selborne on 8 July 1776, Grimm stayed 28 days, for which he was able to charge a significant sum of 2½ guineas per week. He produced twelve views. White described how: 'He first of all sketches his scapes with a lead-pencil; then he pens them all over, as he calls it, with Indian-ink, rubbing out the superfluous pencil-strokes; then he gives a charming shading with a brush dipped in Indian ink; and last he throws a light tinge of watercolours over the whole.'[37] Six of Grimm's drawings were translated for the first edition of the book by professional engravers including William Angus, Daniel Lerpinière and Peter Mazell. These included an expansive topographical view of the village entitled *North East View of Selborne, from the Short Lythe* (fig.14), which unfolds out from the front of the first edition. It establishes a picturesque conceptual framework of observation from the opening spread of the book.

White would have been conscious of contemporary concepts of the picturesque, which had been defined by the Rev. William Gilpin

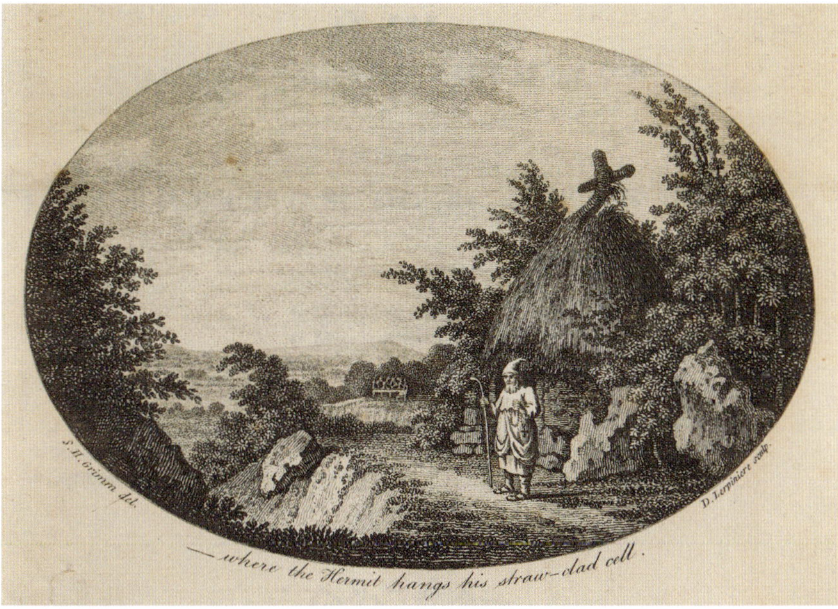

FIG.15
Samuel Hieronymus Grimm, '*Where the Hermit hangs his straw clad cell*'. Etching on paper, engraved by Daniel Lerpinière. Illustration to the first edition of Gilbert White, *The Natural History and Antiquities of Selborne, in the County of Southampton with Engravings, and an Appendix*, London: B. White & Co., 1789.

as 'a term expressive of that peculiar kind of beauty, which is agreeable in a picture' in his *Essay on Prints* (1768), where he sought to outline 'the Principles of picturesque Beauty, the Different Kinds of Prints, and the Characters of the most noted Masters'.[38] In the creation of his garden at The Wakes, White had been influenced by landscape designers such as William Kent to include a ha-ha to keep livestock out without spoiling the view, viewpoints from various urns, and a statue (in fact a flat wooden cut-out rather than the kind of antique stone sculpture that might be seen on a landed estate). In Grimm's image we see a party of figures looking out from the hilltop towards the church and village, with a view towards the woods on the hill beyond. This was known as the Hanger, where White had a Gothic summerhouse for entertaining (and serving his home-grown melons) at the top of a zig-zag path that he and his brother John had cut into the hillside. The summerhouse was a wattled hut with a thatched roof. Known as the Hermitage, it featured in an oval half-title page illustration (fig.15) where it is described underneath as being 'where the Hermit hangs his straw-

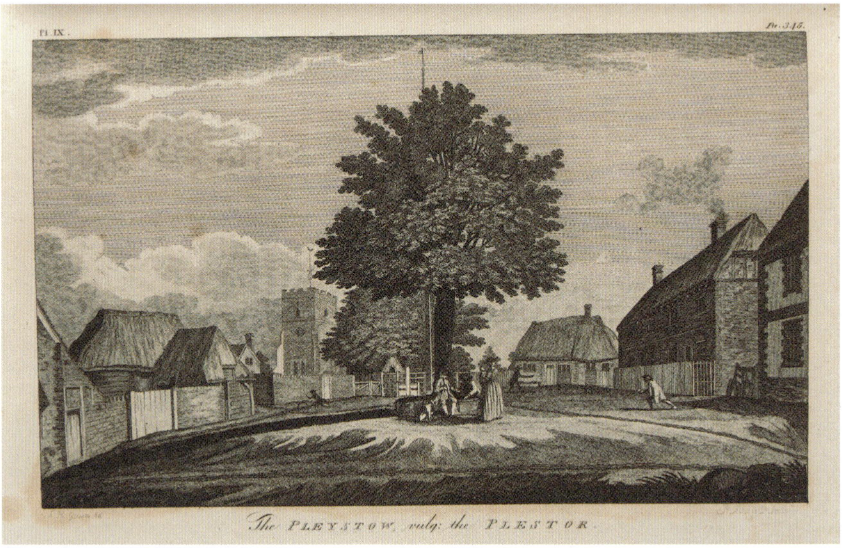

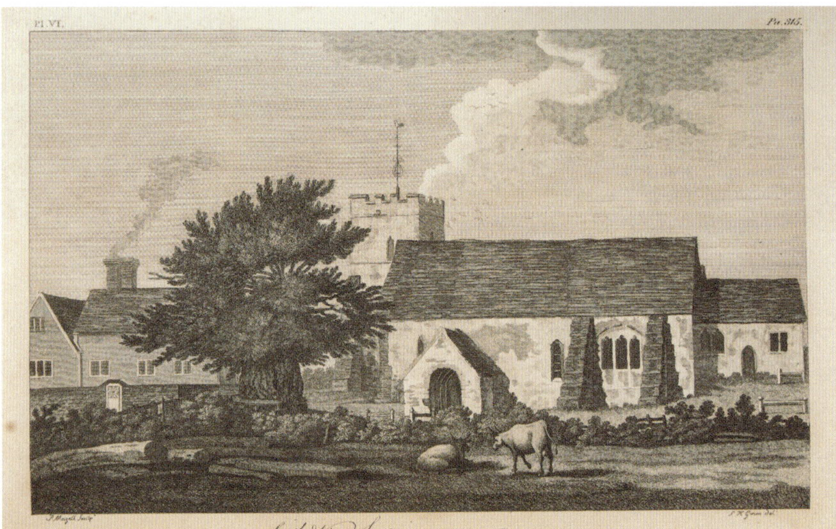

FIG. 16 (above)
Samuel Hieronymus Grimm, *The Plestor.* Etching on paper, engraved by Peter Mazell. Illustration to the first edition of Gilbert White, *The Natural History and Antiquities of Selborne, in the County of Southampton with Engravings, and an Appendix,* London: B. White & Co., 1789.

FIG. 17 (below)
Samuel Hieronymus Grimm, *South View of Selborne Church.* Etching on paper, engraved by Peter Mazell. Illustration to the first edition of Gilbert White, *The Natural History and Antiquities of Selborne, in the County of Southampton with Engravings, and an Appendix,* London: B. White & Co., 1789.

clad cell', a quotation from White's poem 'The Invitation to Selborne' (see p.17).[39] In the eighteenth century, garden hermits or ornamental hermits were invited to live in purpose-built follies or 'hermitages' on the estates of wealthy land-owners, where they would be encouraged to dress like druids and be cared for and consulted for advice, or viewed for entertainment – but, in this case, it was actually White's younger brother Henry dressed 'to great advantage' to look picturesque.[40]

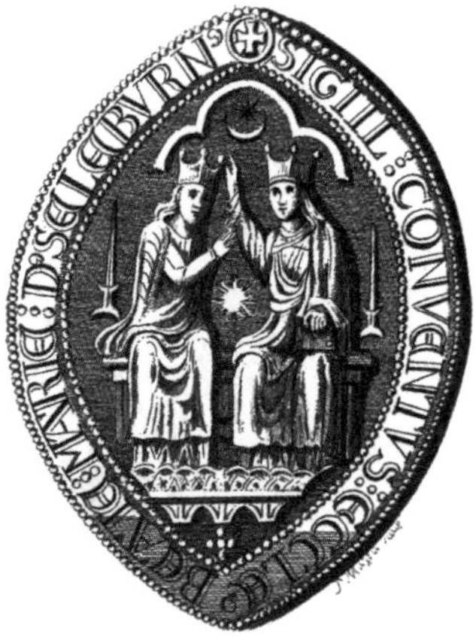

FIG.18
Peter Mazell, *Regina Coeli – Seal of Selborne Priory*. Etching on paper. Illustration to the first edition of Gilbert White, *The Natural History and Antiquities of Selborne, in the County of Southampton with Engravings, and an Appendix*, London: B. White & Co., 1789.

Gilbert White's friend Thomas Mulso was sceptical on seeing a proof of the hermitage engraving: 'I declare that had the Picture come thro' any Hands but *a White's*, which might have directed me, I should not have guessed at the Place. A Print in general does ill with Perspective; but in this, neither the Hill itself, or the neighbouring Country are in Character.'[41] Other illustrations in the first edition included more prosaic views of the ancient manor house (*Temple House, Selborne*); the village green known as the Plestor (from the Saxon words *pleg stow*, meaning *play space*; fig.16); views of Selborne Church from the south (fig.17) and the north; and a drawing of the seal of Selborne Priory (also engraved by Mazell; fig.18).

With a nod to geology, the book includes an engraving of fossil shells of the cockscomb oyster described by White as *Mytillus, Crista Galli* (fig.19). This accompanies White's third letter to Thomas Pennant:

> The fossil shells of this district, and sorts of stone, such as have fallen within my observation, must not be passed over in silence. And first I must mention, as a great curiosity, a specimen that was ploughed up in the chalky fields, near the side of the Down, and given to me for the singularity of its appearance, which, to an incurious eye, seems like a petrified fish of about four inches long, the cardo passing for a head and mouth. It is in reality a bivalve of the Linnean Genus of Mytilus, and the species of Crista Galli.[42]

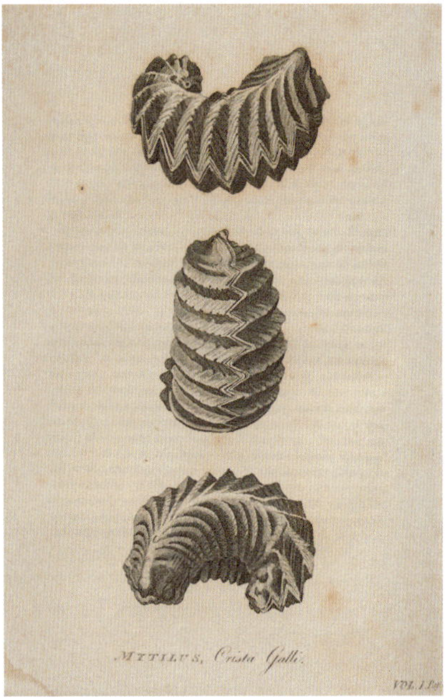

FIG. 19
'Mytilus, Crista Galli – a fossil'. Etching on paper. Illustration to vol. I, page 12 of the first edition of Gilbert White, *The Natural History and Antiquities of Selborne, in the County of Southampton with Engravings, and an Appendix,* London: B. White & Co., 1789.

White recounts how he had sought examples of the fossil in London, but never found an entire specimen, nor had he found in any books 'any engraving from a perfect one'.[43] He proceeds to describe it, but ultimately admits his belief in the primacy of art over language in this particular case: 'the curious foldings of the suture the one into the other, the alternate flutings or grooves, and the curved form of my specimen being much easier expressed by the pencil than by words, I have caused it to be drawn and engraved.'[44]

However, the artistry of others was not always as reliable as White's own written descriptions. A hand-coloured copper engraving *A Hybrid Bird* (fig.20) was accompanied by the explanation: 'Mr Elmer of Farnham, the famous game painter, was employed to take an exact copy of this curious bird.'[45] Stephen Elmer ARA (1717–96) was an eighteenth-century painter who had a reputation for depicting still life and dead game, and who exhibited at the Royal Academy of Arts from 1772 until his death. His picture depicts a hybrid pheasant that White had been sent by Lord Stawell from the great lodge in Holt after it had been found by one of his keepers' spaniels. White's description is precise: 'the shape, and air, and habit of the bird, and the scarlet ring round the eyes, agreed well with the appearance of a cock pheasant: but then the head and neck, and breast and belly, were of a glossy black… The legs and feet were naked of feathers; and therefore it could be nothing of the grous kind.'[46] However, even if the drawing and White's description were accurate, the person who hand-coloured Miller's engraving after Elmer's painting did not pay attention to the author's detailed and specific description of the bird's colouring. Edward Turner Bennett, editor of the 1837 edition, noted that he had examined

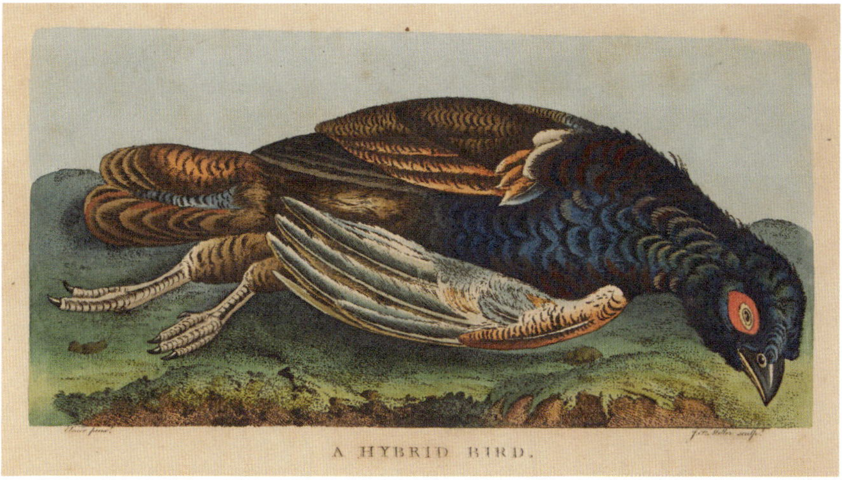

FIG.20
After Stephen Elmer ARA, *A Hybrid Bird*. Hand-coloured etching on paper, engraved by John Frederick Miller. Illustration to Gilbert White, *The Natural History and Antiquities of Selborne, in the County of Southampton with Engravings, and an Appendix* London: White, Cochrane & Co; Longman, Hurst, Rees, Orme, and Brown; J. Mawman; S. Bagster; J. and A. Arch; J. Hatchard; R. Baldwin; and T. Hamilton, 1813.

the same 'curious bird stuffed in the collection of the Earl of Egremont at Petworth, in the year 1804'. He agreed White's description 'to be quite correct,' but noted that 'the plate was exceedingly ill coloured, which indeed may be perceived by comparing it with the description.'[47]

Many of the Victorian editions feature steel- or wood-engravings in the style of well-known artists of the time, such as rural landscapes reminiscent of Joseph Mallord William Turner (1775–1851) or the country cottages of Myles Birket Foster (1825–99). Of the contemporaneous artists associated with Gilbert White, the Northumbrian wood-engraver and natural history author Thomas Bewick (1753–1828) is by far the most celebrated, despite never actually producing any illustrations for *Selborne* himself, and never corresponding directly. Bewick was some thirty years younger than White, but the parson–naturalist and the artist–naturalist shared an affinity of approach. Both were recorders of the minutiae of country life and precise observers of birds and animals. By all accounts, Bewick admired White's writings. His own books – *A History of Quadrupeds* (1790) and the two-volume *A History of British Birds* (the first volume of which was published in 1797) – reflected the vogue for natural history volumes which followed the publication of White's book. In the first volume of the

British birds book, dedicated to land birds, Bewick presented a list of nineteen birds which were 'chiefly selected from Mr. White's *Natural History of Selborne*, and are arranged nearly in the order of their appearing'.[48] The wryneck ('Middle of March') begins the list, the nightingale is placed at the beginning of April, the cuckoo in the middle of April, and it ends with the flycatcher in the middle of May.[49]

Both Bewick and White expressed poetry in the particular. In the opening words to his 1942 book *British Romantic Artists,* John Piper cited 'the particularization of Bewick about a bird's wing', seeing it as 'the result of a vision that can see in these things something significant beyond ordinary significance: something that for a moment seems to contain the whole world; and when the moment is past carries over some comment on life or experience besides the comment on appearances.'[50] When Piper was courting Myfanwy Evans in the early 1930s, he gave her a copy of White's *Selborne* featuring illustrations that were listed as being by Bewick.[51] In fact, several of the vignettes that were originally engraved for Bewick's *A History of British Birds* had been copied as headpieces and tailpieces for nineteenth-century editions of *Selborne*. Bewick's daughters were understandably greatly annoyed, but although they complained to the publishers, several later editions were still issued with the claim that Bewick was the illustrator, as his name carried huge public appeal.[52] In the Harting edition of 1884, a version of Bewick's striking image of a raven accompanies the story of the felling of an oak tree which was home to a pair of the birds, so that it was 'distinguished by the title of the Raven-tree' (fig.21a),[53] and a chaffinch perched on a branch accompanies an account of vast flocks appearing in the fields in January 1768 (fig.21c). A stately group of peacocks from Bewick's *History of British Birds, vol I* sits alongside White's observation of the 'magnificent birds' at the home of his neighbour (fig.21f).[54] Images in a complementary style were contributed by Bewick's pupils, including John Thompson (1785–1866) who illustrated Lady Drover's edition of 1833 with images of birds such as a fern-owl (or night-jar) eating a fly (fig.59); John Jackson (1801–48) who contributed romantic topographical images such as a view of The Wakes (fig.22); Charlton Nesbit (1735–1838) who illustrated the 1836 edition; and Charles Gray (–1845). It is perhaps understandable that Bewick and White have become so intertwined, despite never meeting or corresponding, as Bewick was to be a strong influence, both directly and indirectly,

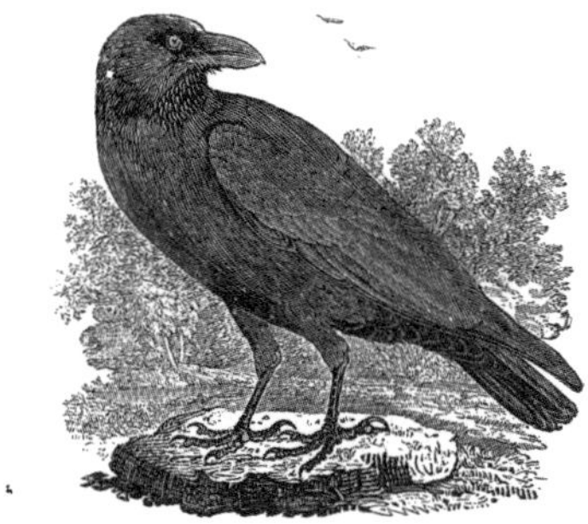

THE RAVEN.

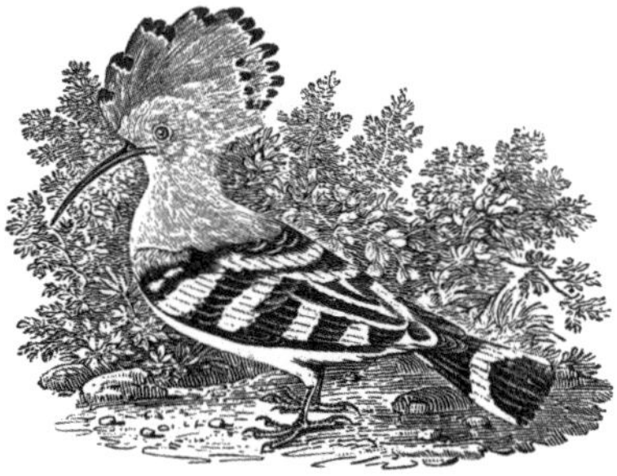

THE HOOPOE.

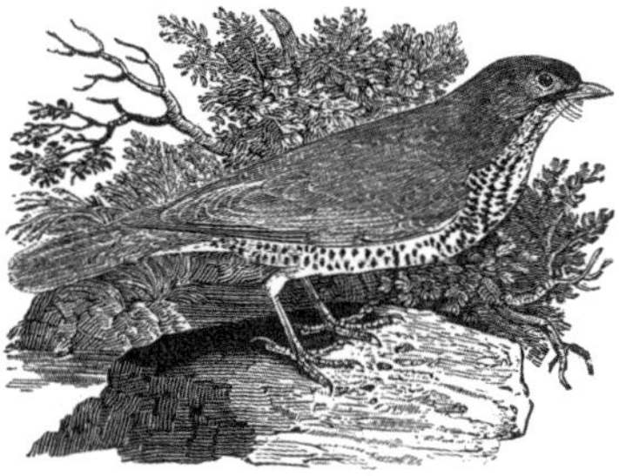

MISSEL-THRUSH.

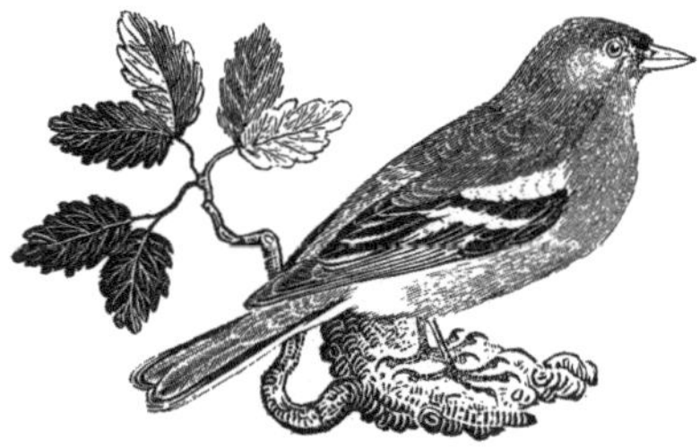

CHAFFINCH.

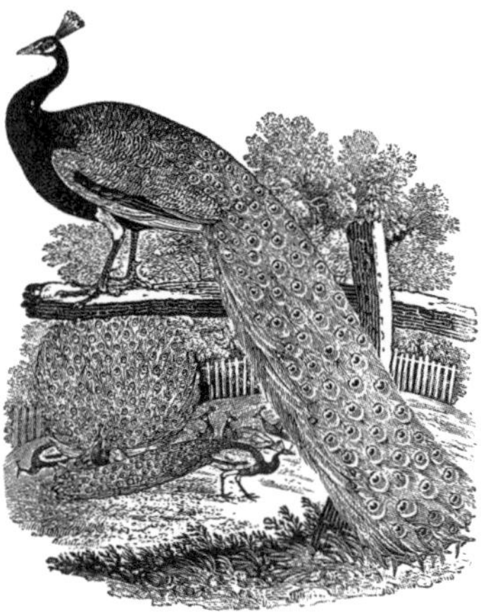

PEACOCK.

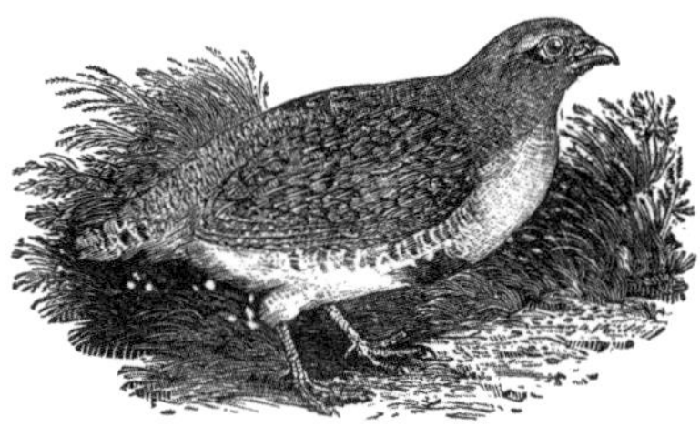

PARTRIDGE.

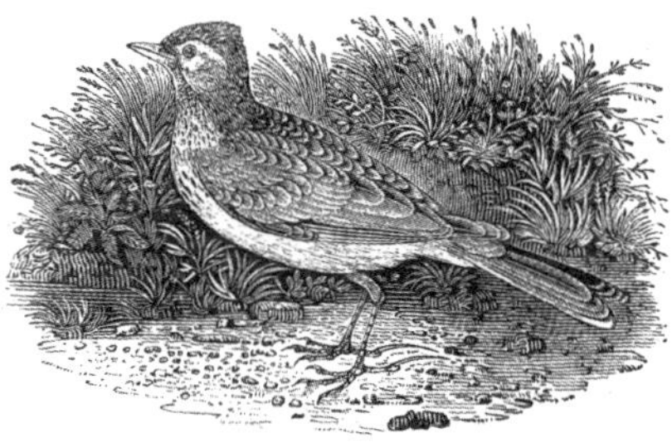

SKYLARK.

FIG. 21a–g
Wood-engravings after Thomas Bewick (clockwise, from top left): *Raven*, *Hoopoe*, *Chaffinch*, *Partridge*, Skylark, *Peacock*, *Missel-Thrush*, c.1797. All: wood-engraving on paper, after *A History of British Birds* reproduced in the James Harting edition of *The Natural History of Selborne*, London: Bickers & Son, 1875.

on so many of the artists who illustrated White's *Natural History* in the nineteenth and twentieth centuries.

The Selborne Society (or Selborne League) was formed in 1885 to 'perpetuate the name and interests of Gilbert White, the Naturalist of Selborne', and reflected the growing interest in environmental issues and the preservation of nature. It was Britain's first national conservation organisation, and within a few years it was publishing the *Selborne Letters* and the *Selborne Magazine*, with the poet Alfred, Lord Tennyson appointed as president in 1888. Members of the Selborne Society were also closely involved with the establishment of the National Trust with Octavia Hill, and the Plumage League, which became the Royal Society for the Protection of Birds (RSPB) in 1904. With the growing interest in White followed numerous new editions of his book. The first of the twentieth century was a handsome Arts and Crafts volume published by the Bodley Head in

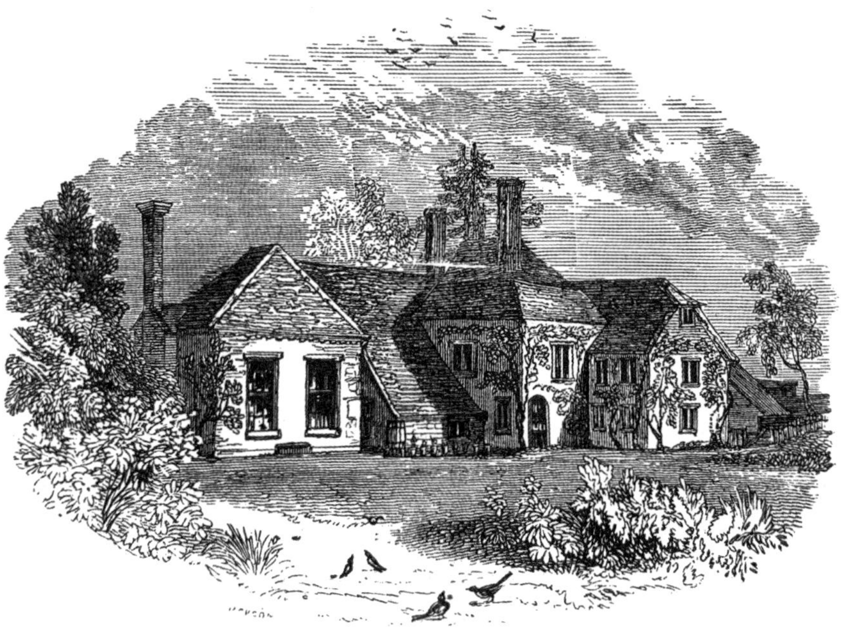

FIG.22
John Jackson, *Back View of the Residence, at Selborne, of the Rev. Gilbert White.* Wood-engraving on paper, from *The History of British Birds* reproduced in the James Harting edition of *The Natural History of Selborne*, London: Bickers & Son, 1875.

1900. Edited by Grant Allen, it featured illustrations by Edmund Hort New (1871–1931), an artist who was a member of the Birmingham Group and who had worked for William Morris' Kelmscott Press. New had illustrated Izaak Walton's *The Compleat Angler* for the Bodley Head in 1895, and the Arts and Crafts stylisation of nature informs the cover design of both books. White and Walton were often to be grouped together as folk-writers, and in the words of J.R. Lowell: 'Nature… endowed these men with the simple skill to make happiness out of cheap material that is within the means of the poorest of us. The good fairy gave them to weave cloth of gold out of straw.'[55]

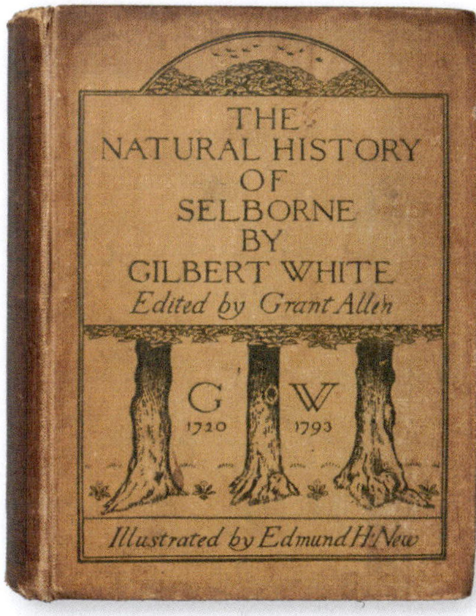

FIG.23
Edmund Hort New, cover of *The Natural History of Selborne by Gilbert White,* Grant Allen, ed. London: The Bodley Head, 1900.

On New's book cover design, the title is borne by an arboretum with swifts (*hirundines*) flying above and the initials and life-dates of Gilbert White placed between the tree-trunks below (fig.23). The opening paragraph of each chapter features a decorative initial letter, a typographic device surrounded by natural forms – leaves, thistles, acorns, birds and flowers (used as the initial letters throughout *this* book). These offer visual clues to signify the beginning of the text, providing emphasis to the letter structure (inevitably the letter 'D' appears frequently, given that most letters start with 'Dear'), but also encapsulating the themes of the text that follows. New's detailed line-drawn illustrations (reproduced through the line-block process) highlight the locations mentioned in the letters, such as particular farms, cottages, country lanes and ponds. Some are based on engravings after Grimm, others on photographs by a 'Professor Bell', as well as interiors of Selborne church and The Wakes, which New had visited in person (fig.24). Within each section are vignettes of animals, birds and insects as they appear: White's thrush, a house-cricket, a hedgehog, and even a drawing of the carapace of Timothy the tortoise in the National History Museum (it had been presented by White's niece). In the same year was published Bowdler Sharpe's

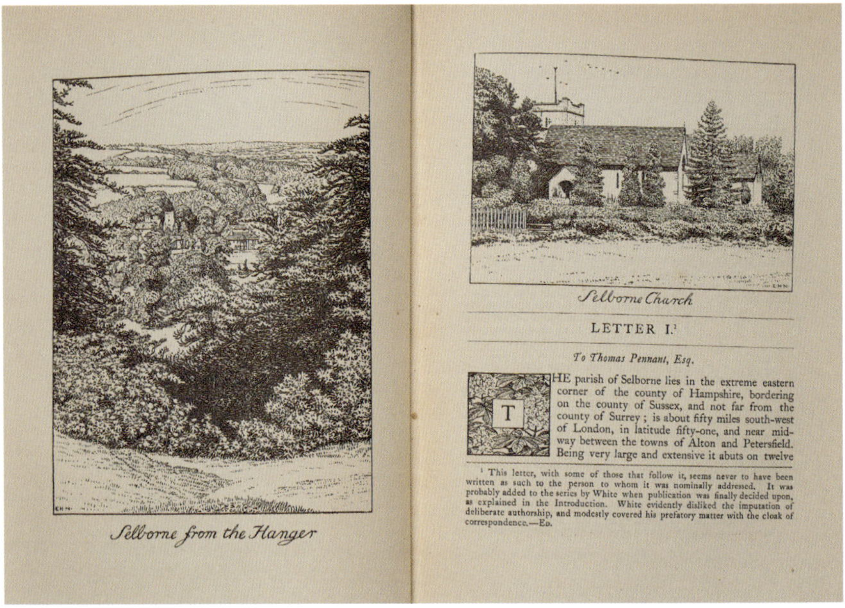

FIG.24
Edmund Hort New, *Selborne from the Hanger* and *Selborne Church*. Illustrations to *The Natural History of Selborne by Gilbert White*, Grant Allen, ed., London: The Bodley Head, 1900.

two-volume edition including the *Garden Kalendar*, with 25 natural history photogravure plates by J.G. Keulemans, topographical drawings by Herbert Railton, and line-drawings by the illustrator Edmund Sullivan (1869–1933), the originals for which are now in the Morgan Library in New York. These included character studies of White as the country squire in his garden – with watering can and sundial (figs 25, 26) – entirely drawn from the imagination, as only two known likenesses were created during his lifetime.[56]

Few of the artists who created illustrations for *The Natural History of Selborne* can have experienced such uncomfortable conditions through their dedication to capturing nature at close hand as the Kearton Brothers, who were among the world's earliest wildlife photographers. In 1895, Cherry Kearton (1871–1940) and Richard Kearton (1862–1928) had published the first natural history book to be entirely illustrated by photographs taken in the wild. Their edition of White's book was first published in 1902, and includes numerous scenes in and around Selborne, as well as many of birds, animals and insects in their habitats. These were hard-won images, requiring

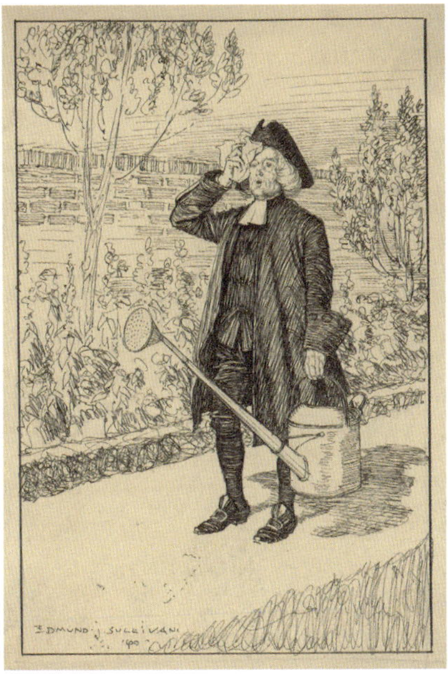

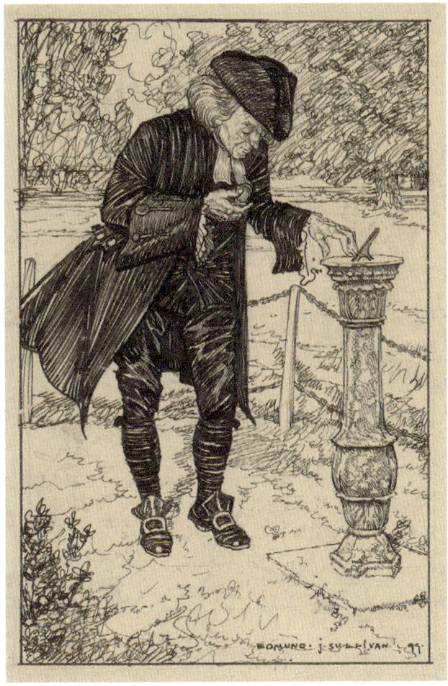

FIGS 25 & 26
Edmund J. Sullivan, *Gilbert White in the Garden* (left) and *The Right Time by the Sun* (right), 1899. Both: ink on paper. Illustration to Gilbert White, *The Natural History and Antiquities of Selborne and A Garden Kalendar*, R. Bowdler Sharp, ed., London: S.T. Freemantle, 1900.

extraordinary camouflage by the photographers, who disguised themselves inside fake livestock and natural forms in order to get close to nature. As Richard Kearton was to muse of White:

> What a letter he would have written to Pennant about his experiences inside our stuffed ox, whilst watching a skylark feed her chicks, and a timid little tree pipit go on to her nest, and the behaviour of wheatears and sandpipers in front of a stuffed sheep with the eye of the camera staring at them from its chest... We wonder what he would have said if he could have crouched with us, in an artificial rock made for this purpose, every day for a week watching at arm's length a pair of ring ouzels feed their young, and waiting until hope almost faded into despair for a gleam of sunshine by which to photograph them.[57]

The Kearton Brothers were the predecessors of modern documentary wildlife photography and filmmaking. The exactness and 'truth' of the relatively new genre of wildlife photography brought with

FIG. 27
Cherry Kearton and Richard Kearton, *Puffins*. Illustration to Letter XXI, 28 November 1768. Photograph, Gilbert White, *The Natural History of Selborne with notes by Richard Kearton*, London: Cassell, 1902.

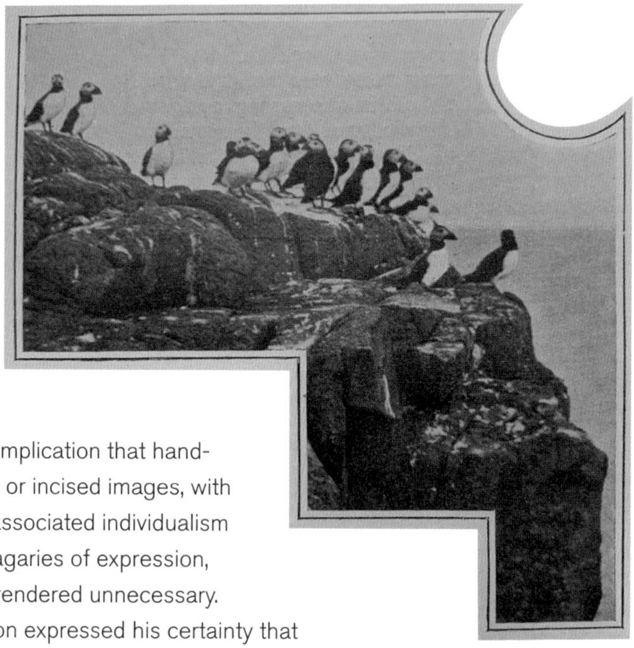

it the implication that hand-drawn or incised images, with their associated individualism and vagaries of expression, were rendered unnecessary. Kearton expressed his certainty that

> We know from his love of truth in illustration, that he would have appreciated the minute accuracy of a good photograph... We believe these to come into closer touch with the spirit of the writer than anything hitherto published. The obtaining of the photographs has been a labour of love. We have given of our best, and whatever may be the judgement of present-day naturalists we feel sure that he, at any rate, would have been the first to recognise our honest endeavour to illustrate worthily that which he gathered together with so much care and pains.[58]

Many of their photographs take the form of framed vignettes, recalling the Bewick format, albeit in a different medium, with others cropped into geometric shapes that create a graphic interplay of text and image (figs 27, 28). At times this cropping has the effect of focussing attention, as if looking through a lens, without the distractions of the surroundings. The Keartons felt that White would never have questioned whether the results were worth the trouble taken and the hardships endured, something they were frequently asked by others. Anticipating how White's vision of rural life has served as a patriotic reminder of a peaceful 'green and pleasant land' at times of conflict, Richard Kearton observed how 'our author's inimitable descriptions of these and a hundred other happenings of the countryside will for all time appeal with irresistible charm to the heart that loves rural England'.[59]

During the First World War, the poet Edward Thomas (1878–1917) wrote *A Literary Pilgrim in England* as a guide to the places that were special to some of Britain's foremost writers, including William Blake, John Keats, Percy Bysshe Shelley, Thomas Hardy, Hilaire Belloc, William Cowper, John Aubrey, Samuel Taylor Coleridge, William Wordsworth and Emily Brontë. Thomas had been living with his wife and young children within ten miles from Selborne at the outbreak of the war, and in his chapter on White he noted how, 'In this present year, 1915, at least, it is hard to find a flaw in the life he led, which we may be excused for looking back dotingly as upon some inaccessible and imperturbable tract of our life'.[60] These words became even more poignant in the light of the poet's death on the Western Front in 1917.

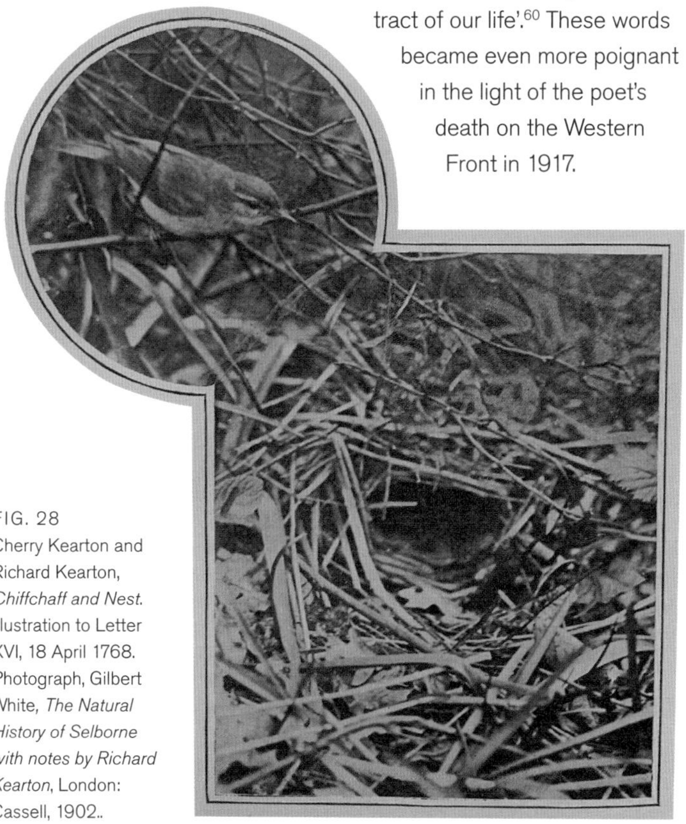

FIG. 28
Cherry Kearton and Richard Kearton, *Chiffchaff and Nest.* Illustration to Letter XVI, 18 April 1768. Photograph, Gilbert White, *The Natural History of Selborne with notes by Richard Kearton*, London: Cassell, 1902..

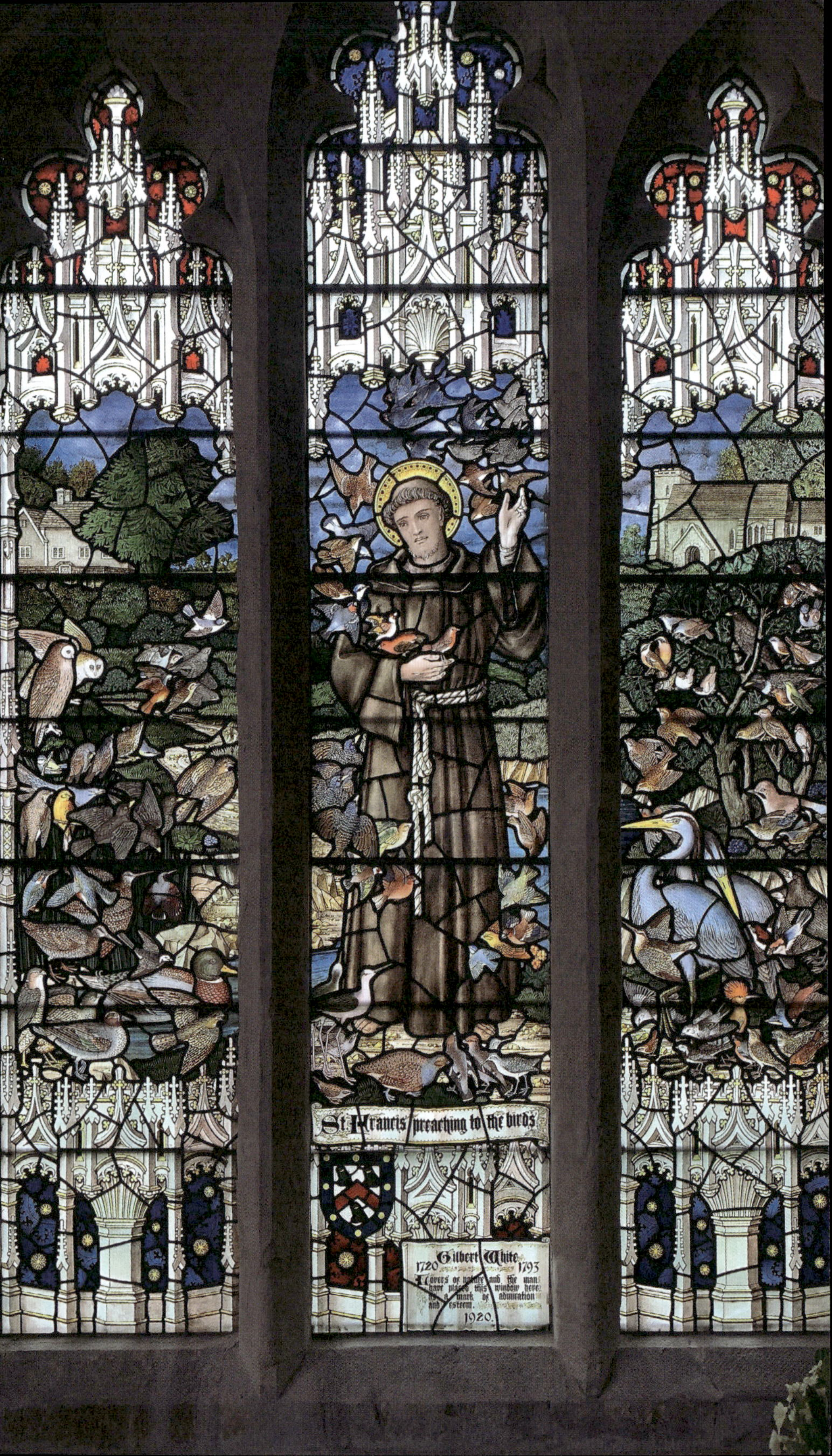
St Francis preaching to the birds
Gilbert White
1720
1793
Lovers of nature and the man have placed this window here as a mark of admiration and esteem.
1920.

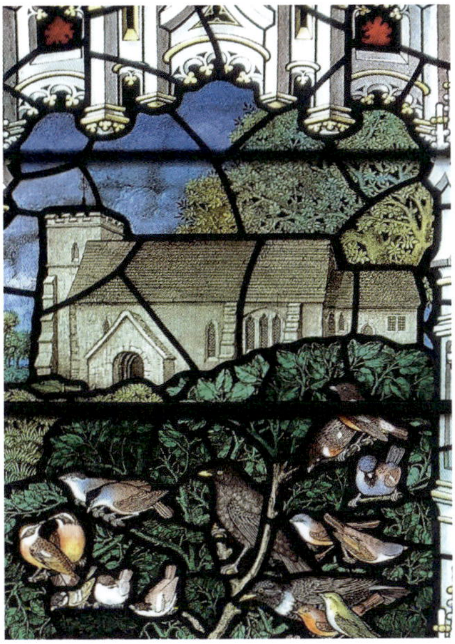

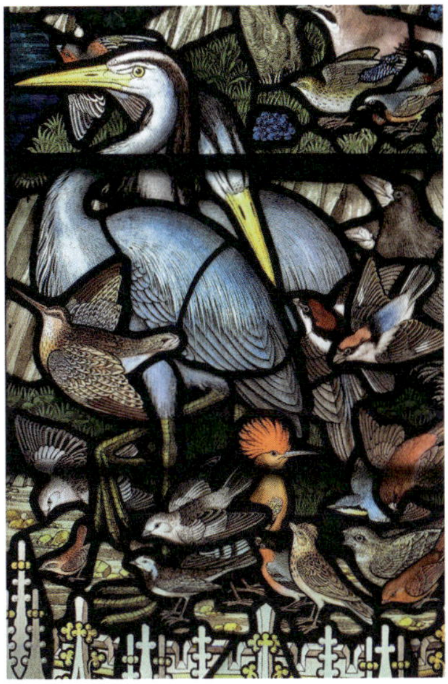

FIG.29 AND DETAILS
Alexander Gascoyne and Horace Hincks of Hinchcliffe, Hincks & Burnell, *St Francis Preaching to the Birds,* 1920. Stained glass in St Mary's Church, Selborne.

In the aftermath of the First World War, to mark the bicentenary of the naturalist's birth in 1920, a stained-glass window was installed in the Church of St Mary in Selborne by the Gilbert White Fellowship (fig.29).. Within three trefoil-arched lights is an astonishing flock of birds, conveying the notion of God as the centre of nature. St Francis of Assisi (*c.*1181–1286) is depicted giving his celebrated sermon to the birds. Transplanted from medieval Italy to Hampshire, the patron saint of animals, birds and ecology comes to represent Gilbert White as the curate of the village. Beyond him is a view of the church itself (based on an engraving after Grimm), the yew tree and the vicarage where White was born. It features over a hundred exquisitely rendered birds, from herons to hoopoes, kingfishers, curlews, wryneck, nightjar and a robin. The window was conceived by the naturalist Joseph Whitaker (1850–1932) who enlisted Alexander Gascoyne (1877–1927) to design it. It was painted under Gascoyne's supervision by Horace Hincks (1886–1964) of Hinchcliffe, Hincks & Burnell, a Nottingham firm of stained-glass designers and manufacturers.[61] Each of the birds depicted is mentioned in White's *Natural History*, and their colouring was taken from two unidentified volumes of *British Birds*.[62]

The window served to reinforce the growing sense of Selborne as a place of pilgrimage for those interested in White and the natural world. In 1926, Mary Kelly staged a pageant in the grounds of The Wakes entitled *The Antiquities of Selborne, Shown to the World by the Rev Gilbert White, of Selborne, in the County of Southampton*, which presented music and scenes from the history of the village to an audience of over 1500, and was restaged in 1938 to an even larger audience.[63] Pageants of this sort promulgated a bucolic vision of the countryside unspoilt by the motorcar and commuter railways (despite an advertising flier for the 1938 presentation stating that 'Electric trains run from London every half hour, leaving Waterloo at 27 and 57 minutes past the hour, arriving at Alton at 49 and 19 minutes past the hour respectively').[64] Paradoxically, in this period Shell Oil was using images of nature to promote the idea of the 'quick starting pair' to represent the reliability and power of using Shell's products simultaneously – an early instance of an advertising technique we might now label 'greenwashing'. In addition to a series of posters featuring pairs of otters, grasshoppers, kingfishers, sea-lions and trout, created between 1926 and 1932, there were memorable posters of an idealised countryside designed by the likes of Paul Nash, Frank Dobson and Graham Sutherland.[65]

A succession of illustrated editions of White's *Selborne* was published during the 1920s and 1930s, when there was a surge of interest in accounts of the countryside, reflecting the wave of rural nostalgia that was manifested in these historical pageants and illustrated country books by the likes of artists John Nash, Agnes Miller Parker and Clare Leighton. There was also a strong revival of

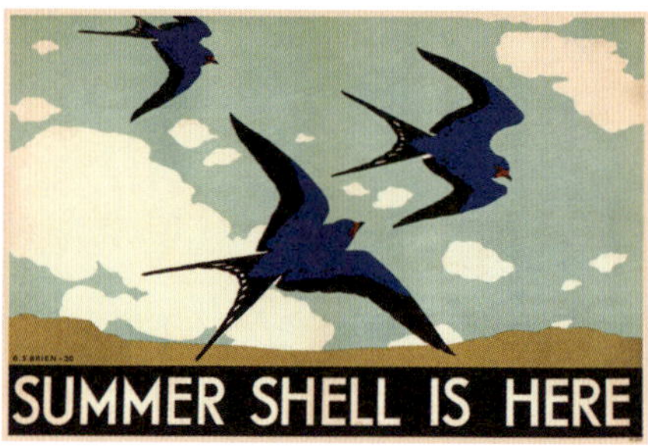

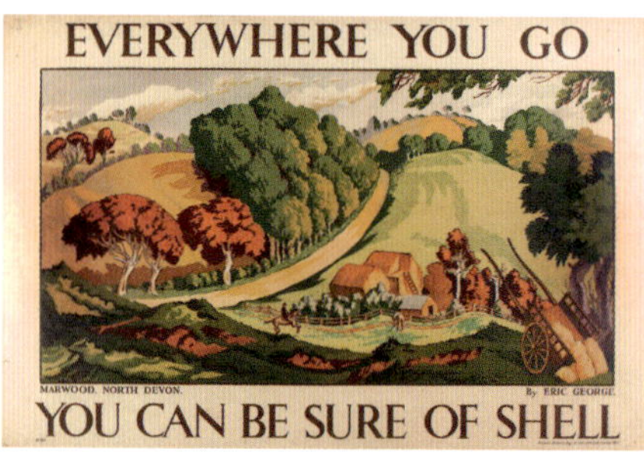

FIG.30
G. Stanislaus Brien, *Summer Shell is Here*, c.1929. Lithograph on paper, Shell Heritage Art Collection.

FIG.31
Eric George, *Marwood, North Devon*, 1932. Lithograph on paper, Shell Heritage Art Collection.

interest in the traditional technique of wood-engraving, reflected in the creation of the Society of Wood Engravers in 1920. For the members an important point of reference was Thomas Bewick, who had popularized illustration in the nineteenth century through his white line engravings and who was widely acknowledged to be the founding father of wood-engraving. Commenting on the popularity of the medium in 1932, Clare Leighton observed:

> In the modern world a continually increasing number of illustrated books are concentrating on wood-engravings. Probably one of the main reasons for this is the mechanical ease with which the engravings can be printed on the same paper as the type, whereas illustrations in so many other mediums have to be printed on separate paper and bound in... Fatigued with the photographic representation of life in the more conventional etching, we call for greater subjectiveness. This is more readily attainable in the wood-engraving than in any other medium.[66]

One of the founder members of the Society of Wood Engravers was the natural history illustrator Eric Fitch Daglish (1882–1966), who had been trained in the medium by the artist Paul Nash. In 1929, Daglish produced a series of images of birds and animals for an edition of White's *Selborne* published by Thornton Butterworth, with an introduction by E.M. Nicholson (1904–2003), a pioneering environmentalist who was later one of the founders of the World Wildlife Fund.[67] Daglish was a professional naturalist and the author of 25 books about the natural world, many of which were illustrated with his woodcuts and engravings, such as *Woodcuts of British Birds* (1925) and a new edition of Walton's *The Compleat Angler* (1927). Daglish's images of wildlife for *Selborne* have an earnest directness, and each is individually printed from the block and tipped into the book. His sixteen *Selborne* prints are entirely engraved with the tint tool, which cuts lines of uniform width with extreme fineness; there is no variety of the thickness of the line, just the closeness of the marks. With this technique he was able to engrave a wide scale of tone in a small area, with lines following the shape of the form. These fine lines are used to extraordinary effect to convey the blades of grass on the turf surrounding a stone curlew (fig.32), the details of the feathers on

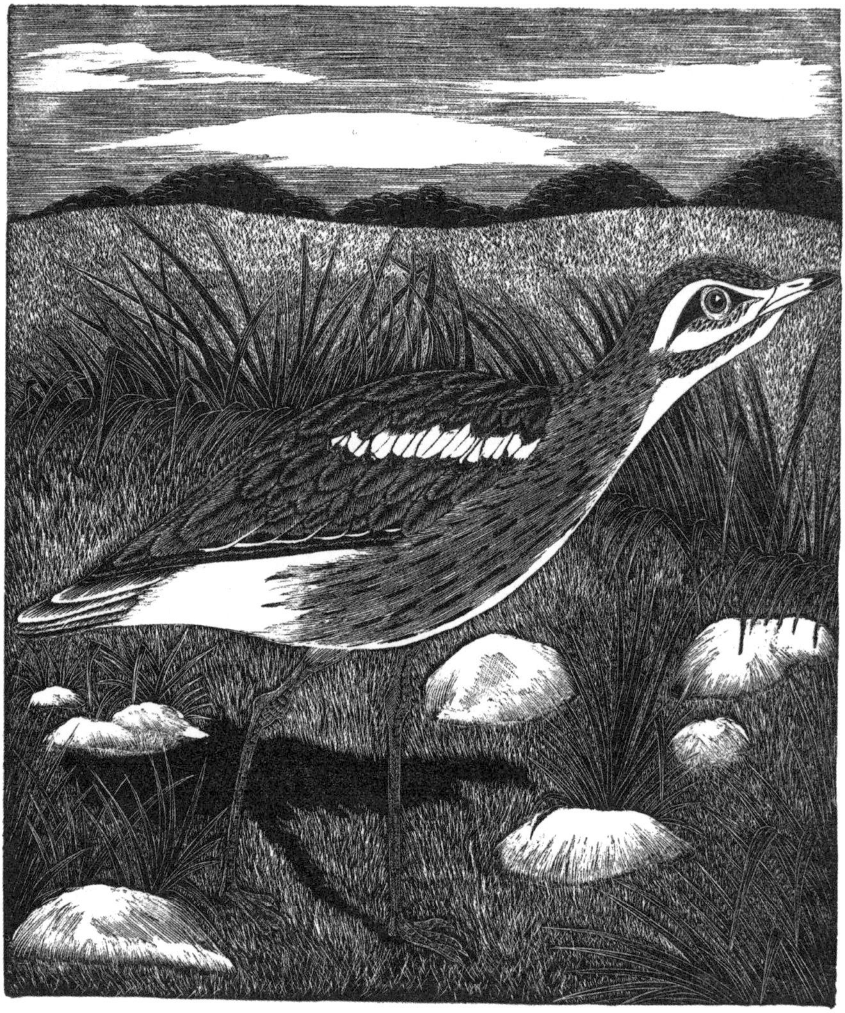

FIG.32
Eric Fitch Daglish, *Stone Curlew,* 1929. Wood-engraving on paper. Illustration to Gilbert White, *The Natural History of Selborne*, E.M. Nicholson, ed., London: Thornton Butterworth Ltd.,1929.

a white (or barn) owl (fig.33), or the individual hairs and whiskers of a snow-bound hare laying sullenly in its seat (fig.34). Such was his interest in the natural world that Daglish subsequently became a lecturer in zoology, later retiring to the country as a field naturalist and correspondent for the BBC.

The sculptor and printmaker Gertrude Hermes (1901–83) was one of the leading women artists in the Society of Wood Engravers. Her prints

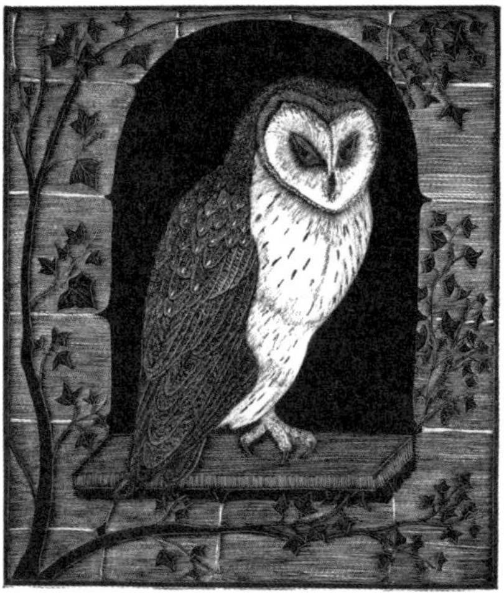 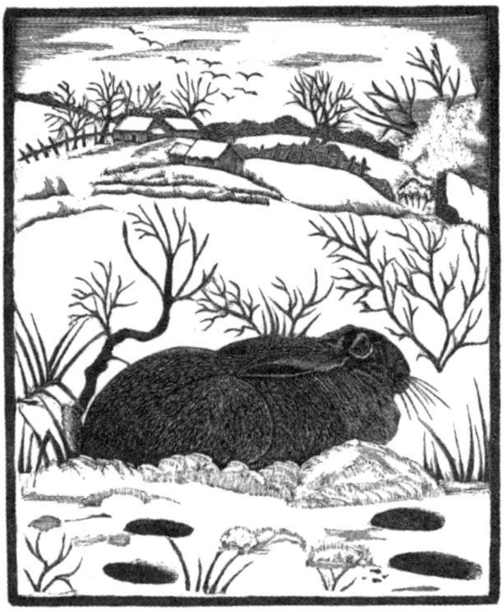

displayed great originality, fusing an awareness of traditional craft with the approach of a sculptor whose modernist aesthetic was comparable with her friends Henry Moore and Leon Underwood. In 1931, Hermes embarked on a group of striking wood-engravings to illustrate an edition of the *Natural History* for the Gregynog Press, a private press in mid-Wales that had been founded in 1922 by the wealthy Davies sisters, who were remarkable collectors and patrons of modern artists. Hermes and her husband Blair Hughes-Stanton had recently arrived in Montgomeryshire to run the press, together with their friends the printer William McCance and his wife the wood-engraver Agnes Miller Parker. It should have been a perfect commission for Hermes, who had a lifelong interest in nature and who created numerous sculptures of animals, birds and reptiles during the 1920 and 1930s, including *Frog* (1926), *Bird in Hand* (1931) and *Butterfly* (1936). As her sister May was to observe: 'We all sensed there was something special about Gertie and her relations with animals, trees, plants, insects. She watched them so intently – seemed more finely tuned to all living things than most of us… Hers was a gentle wildness.'[68]

FIG.33
Eric Fitch Daglish, *White (or Barn) Owl,* 1929. Wood-engraving on paper. Illustration to Gilbert White, *The Natural History of Selborne*, E.M. Nicholson, ed., London: Thornton Butterworth Ltd., 1929.

FIG.34
Eric Fitch Daglish, *Hare Snow-bound,* 1929. Wood-engraving on paper. Illustration to Gilbert White, *The Natural History of Selborne*, E.M. Nicholson, ed., London: Thornton Butterworth Ltd., 1929.

Whilst working on the commission, Hermes lodged in Selborne for a week with the village postman and his wife in order to absorb

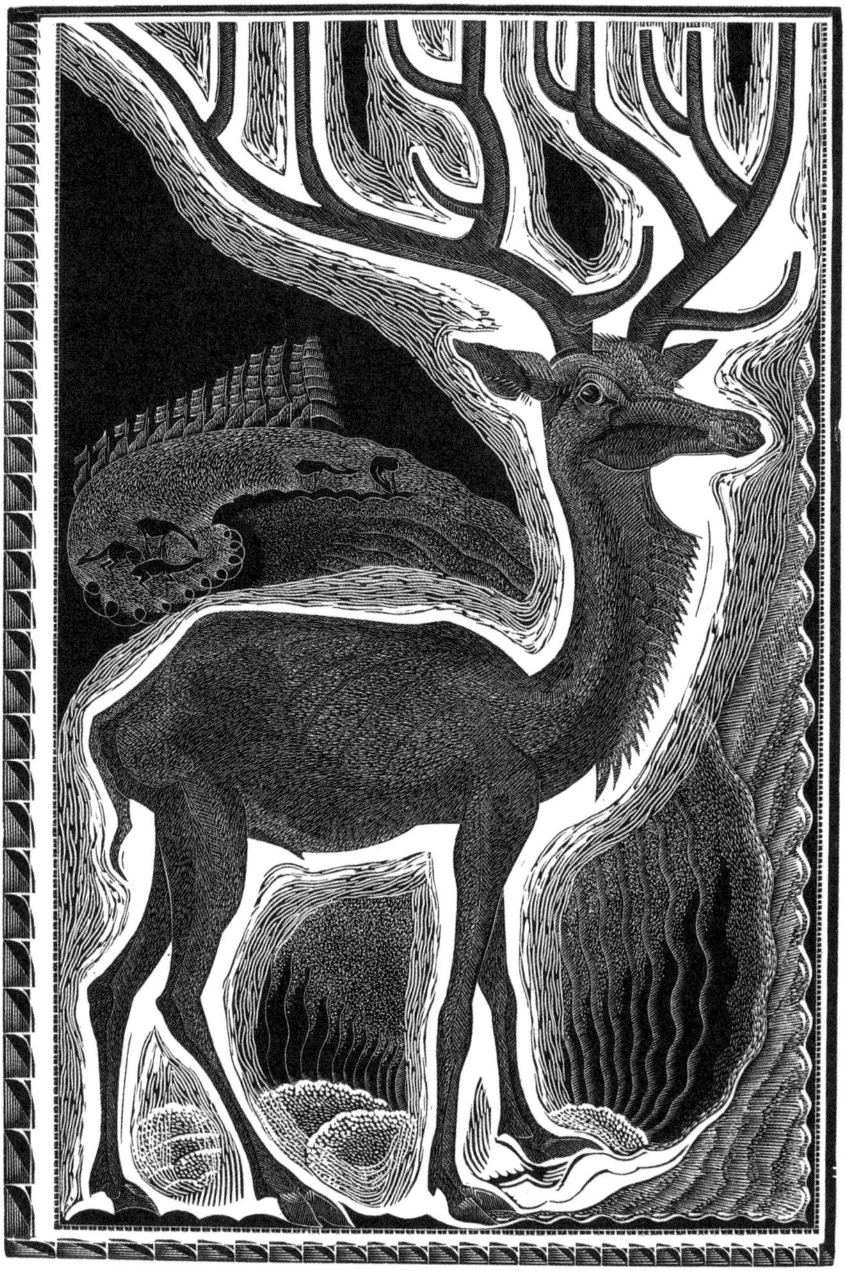

FIG.35
Gertrude Hermes, *Deer*, 1932. Wood-engraving on paper, in William Condry, *Wood Engravings by Gertrude Hermes being Illustrations to Selborne with Extracts from Gilbert White,* Newtown: Gwasg Gregynog, 1988.

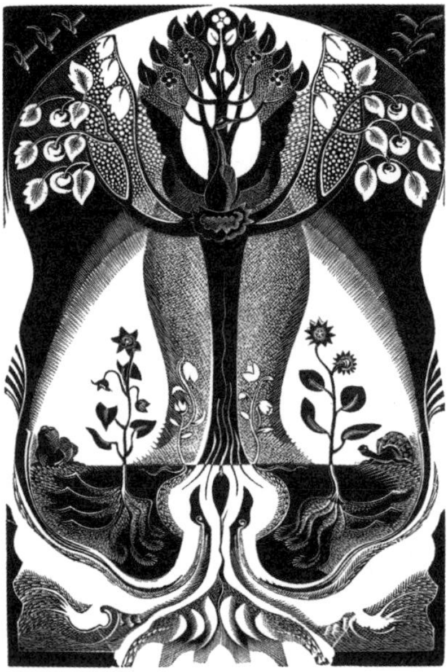

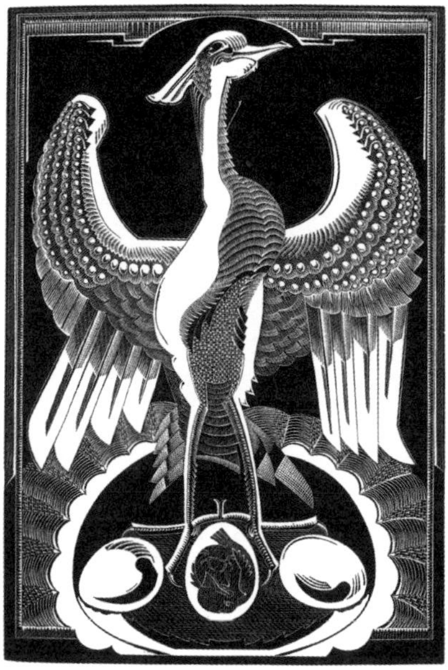

the sense of place and spirit of White's writings, becoming such good friends with them that she continued to order Cox's Orange Pippins and honey from them for the next two or three years. The resulting set of six engravings stand alone as some of the most distinctive and singular images in response to White's writings. The wood-engraver Simon Brett has described them as 'perhaps the most refined, the most exquisite wood-engravings she made'.[69] *Tree* (fig.36), an image with roots stretching into the earth and branches extending into the sky, brings to mind Rorschach blots or Surrealist associational images – it is at once suggestive of a tree, an apple core and human sexual reproductive organs. It conveys the fecundity and interrelatedness of nature, of sky and earth.

FIG.36
Gertrude Hermes, *Tree*, 1932. Wood-engraving on paper, in William Condry, *Wood Engravings by Gertrude Hermes being Illustrations to Selborne with Extracts from Gilbert White*, Newtown: Gwasg Gregynog, 1988.

FIG.37
Gertrude Hermes, *Heron*, 1932. Wood-engraving on paper, in William Condry, *Wood Engravings by Gertrude Hermes being Illustrations to Selborne with Extracts from Gilbert White*, Newtown: Gwasg Gregynog, 1988.

Sending the block for the engraving *Heron* (fig.37) to McCance at Gregynog in June 1932, the exacting artist wrote: 'It's the 'Hermes Bird' – never been seen at Selborne! There is a hopeless slip on the neck… but really that was the worst block I have ever worked on – dreadfully uneven in quality.'[70] In spite of this, Hermes was

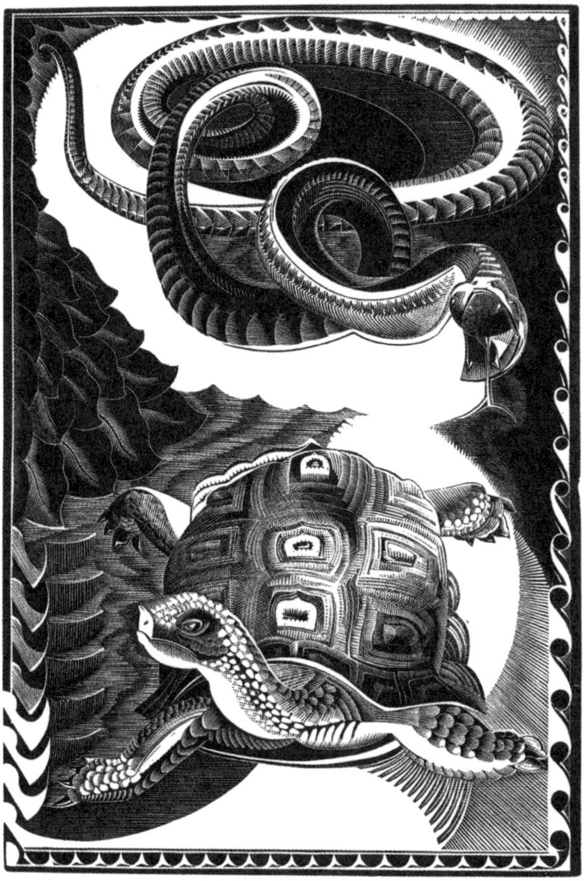

FIG.38
Gertrude Hermes, *Snake and Tortoise*, 1932. Wood-engraving on paper, in William Condry, *Wood Engravings by Gertrude Hermes being Illustrations to Selborne with Extracts from Gilbert White,* Newtown: Gwasg Gregynog, 1988.

undoubtedly immensely skilled and controlled. These are a sculptor's prints, revealing an interest in the formal qualities of the subjects depicted and the interplay of positive and negative space. McCance wrote to Hermes, 'Your scheme is admirable to do the full pages of decorative designs each of a different genre, reptile, birds, flowers, beasts, insects.'[71] Amongst the group is *Carp* (fig.84), an underwater scene with carp, newts and water weed, and *Spider and Bat* (fig.85), a nocturnal scene of a bat hanging upside down above a spider – with the stars in the night sky twinkling through a cobweb that has entrapped flies that wait to be consumed. The cultural historian Alexandra Harris has likened these prints to 'prehistoric fossils', noting of the heraldic image of a majestic stag with its antlers extending upwards like tree-branches (fig.35): 'This stag stops apprehensively for a moment and catches our eye before leaping away. Yet he seems also to be permanent, representing Selborne for all time.'[72]

The engravings bear an astonishingly complex array of marks and patterned borders. Hermes spoke of herself as a book 'decorator' rather than a book illustrator, yet it would be difficult to describe these as 'decorative' images in the conventional sense. There is a subtle air of menace. In an unsettling pairing of two reptiles – *Snake and Tortoise* (fig.38) – Hermes demonstrates her artistic virtuosity, with quite literally serpentine cross-hatching animating the writhing body of the snake. As Brett has observed, 'Behind the tortoise's head, tints delicately laid across the shell set the tones of the head into relief; while the head and the rear leg, just above it, have planes angled into pure white, a sculptor's decision which produces bold shapes for the pattern-maker.'[73]

Although these are some of Hermes' finest prints, the strength of her subjective vision is perhaps out of kilter with White's gentle and often humorous observations, and the prints were not published during Hermes' lifetime. Whilst working on them she had experienced a period of great unhappiness relating to the break-up of her marriage to Hughes-Stanton, which she described as 'a distressing cycle of events for me, quite unforeseen and utterly nerve-shattering',[74] and, despite her willingness to continue, the directors of the Gregynog Press decided that 'the book be postponed until a more favourable time'.[75] This was not to be until 1988, when the engravings were published in a book in a limited edition of 200 copies printed from the original blocks by the Gregynog Press to mark the bicentenary of White's *Natural History*.[76]

Although Gertrudes Hermes' prints were not published during the 1930s, at this time there was no shortage of other Selborne-related books, including Gilbert White's journals, edited by the nature writer Walter Johnson and published in 1931. The reviewer in *The Spectator* noted that:

> It is hard to believe, in a bookmaking age among a people possessed by a passion for natural history, that the diaries of Gilbert White have lain all these years in a museum unread, unsung. He was among the most English of Englishmen, the father of a cult that spreads around the world… Who wishes to know England and the English cannot omit this record.[77]

Linking White to questions of Englishness was particularly apt, as rural England was then perceived as being under threat from development and change. The writer on ruralism H.J. Massingham (1888–1952) had written a number of nostalgic country books that were influenced by White, including *Downland Man* (1926), *Country* (1934) and *The Genius of England* (1937), and he edited a two-volume edition of *The Writings of Gilbert White*, published by Nonesuch Press in 1938.[78] In his introduction he declared a desire to present White as 'the great countryman rather than the great naturalist'.[79] This ambition was reflected in the illustrations that accompanied the text by Eric Ravilious (1903–42), whose wood-engravings have a greater focus on rural scenes of country life 'rather than careful pictures of the birds and flowers that White mentions by name.'[80] Like others before him, Ravilious went to Selborne and walked in the landscape. There are views of *Woolmer Forest* (fig.97), an owl flying down a hollow lane (fig.71), and others based on Grimm's engravings, such as a reimagining of the 'Hermit' at his picturesque abode (fig.98) and *The Plestor* – the enclosed grassy area where villagers gathered around a vast oak (fig.93). Ravilious presents a nostalgic view of Selborne past that avoids the present, which is perhaps understandable given his comment to the painter Diana Low of how, 'It is a book I wouldn't care for with pictures to tell the truth – at least not any contemporary.'[81] The title pages for both volumes show views looking from interiors out to nature. One is a view towards the church (fig.9). White had written that church towers 'are very necessary ingredients in an elegant landscape'.[82] The other (fig.89) looks down the nave of the church into an ordered garden with a greenhouse, a subject of special appeal to the artist.[83] Both include emblematic depictions of White (volume one with his friend and correspondent Thomas Pennant, and in volume two alone) that are reflective of popular Georgian silhouettes. The title page for volume one seems to compress eleven of Bewick's birds, plus White's celebrated tortoise Timothy, into one stylized summary image. It sets up the convivial tone of the book with nature pressing in on all sides.

Ravilious had loved both Bewick's wood-engravings and White's *Selborne* since his student days. In the 1920s, he encouraged his own students at Eastbourne College of Art so that 'all were soon reading *The Natural History of Selborne*'.[84] He admired White's poetic turn-of-phrase, and wrote to his friend Helen Binyon in January

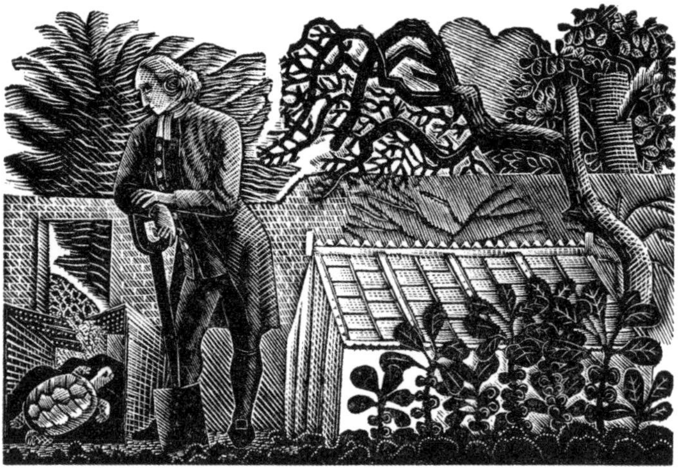

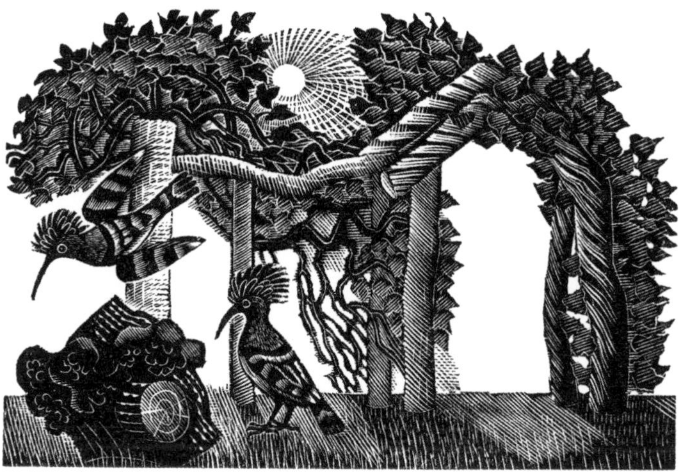

1936: 'The *History of Selborne* is a grand book and I read it every minute I can spare from engraving and other jobs. – "There are bustards on the wide downs near Brighthelmstone." Isn't that a beautiful statement?'[85] The pairing of his headpieces with White's texts is immensely pleasing. The black and white medium might perhaps seem inappropriate for illustrating White's evocative description of the noonday sun appearing 'as blank as a clouded moon' and creating 'a rust-coloured ferruginous light' that signalled

FIGS 39 & 40
Eric Ravilious, *The Tortoise in the Kitchen Garden* (above) and *The Hoopoes* (below), 1938. Both: wood-engraving on paper. Illustration to *The Writings of Gilbert White – Selected and Edited with an Introduction by H.J. Massingham*, London: The Nonesuch Press, 1938.

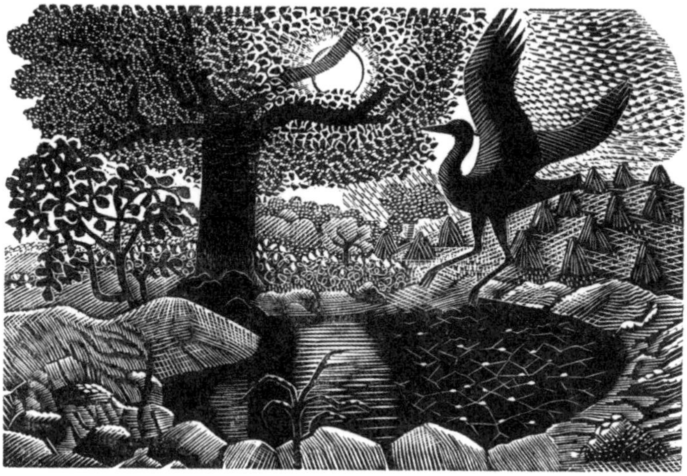

FIGS 41 & 42
Eric Ravilious, '*A rust-coloured ferruginous light*' (above), *Feeding the Fowls* (opposite), 1938. Both: wood-engraving on paper. Illustration to *The Writings of Gilbert White – Selected and Edited with an Introduction by H.J. Massingham*, London: The Nonesuch Press, 1938.

the 'amazing and portentous' summer of 1783 (which we now know was the result of the eruption of the Laki volcano in Iceland between 8 June 1783 and February 1784). Yet, Ravilious manages to perfectly convey the hazy sunlight through the leaves of a tree as a heron rises above a pond (fig.41).[86] Somehow, colour is not necessary, just as it is entirely absent in the whiteness of the wintery image to accompany White's account of 'considerable falls of snow, which lay deep and uniform on the ground without any drifting, wrapping up the more humble vegetation in perfect security (fig.91)'.[87] Elsewhere, to accompany White's observation that, 'A good ornithologist should be able to distinguish birds by their air as well as by their colours and shape; on the ground as well as on the wing, and in the bush as well as in the hand',[88] Ravilious presents a joyful flock of birds around a tree, some airborne, others on the ground, or in the tree (fig.90).

'It's quite the best book I've ever been given to do and I love doing it', Ravilious wrote to Diana Low and her husband Clissold Tuely.[89] However, he also complained that engraving over 30 blocks was 'mortal slow work' and it was to be his last illustrated book.[90] There are ample amusing strange moments, such as the hoopoe birds described as marching about 'in a stately manner' in the garden of The Wakes (fig.40);[91] the depiction of White attempting to measure the Duke of Richmond's dead moose in the greenhouse

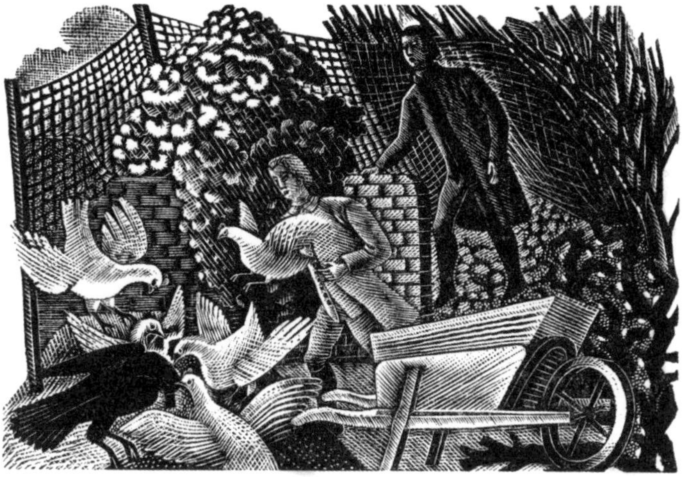

at Goodwood (fig.10), or a gentle image of him leaning on a garden spade watching the movements of Timothy the Tortoise (fig.39). Massingham wrote to the production manager Harry Carter, 'I have shown everybody who comes to see me the Ravilious wood-engravings and they are unanimous in being enchanted with them'.[92] However, they are not entirely joy-filled images. One scene, known as *Feeding the Fowls* (fig.42), has sometimes been misinterpreted in compendiums of Ravilious' work as just a charming image of a young boy feeding the birds – when in fact it is an illustration of the story of nature exacting revenge on a sparrowhawk that had taken most of a neighbour's chickens, showing the carnage that ensued after the man and boy set the hawk down among the brooding hens, having clipped its wings, cut its talons and placed a cork on its beak. Ravilious couldn't resist the challenge set in the text, although the common misinterpretation suggests that perhaps the naturalist was right: 'Imagination cannot paint the scene that ensued; the expressions that fear, rage, and revenge inspired were new, or at least such as had been unnoticed before: the exasperated matrons unbraided, they execrated, they insulted, they triumphed. In a word, they never desisted from buffeting their adversary till they had torn him in a hundred pieces.'[93]

During the Second World War, the journalist and satirist Leonard Russell started *The Saturday Book,* an annual miscellany featuring poetry and essays on a range of arts. He declared of the anthology's

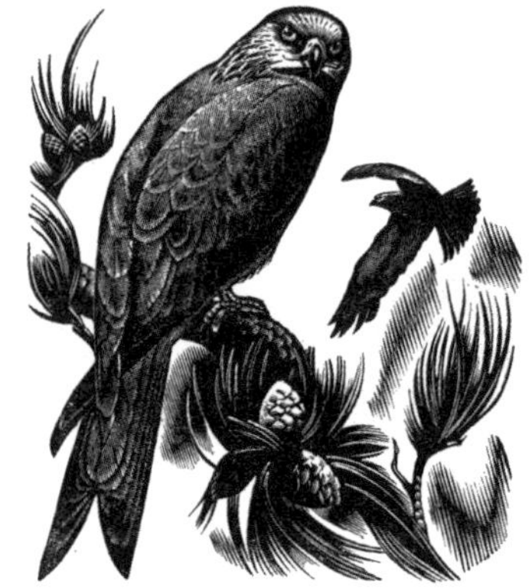

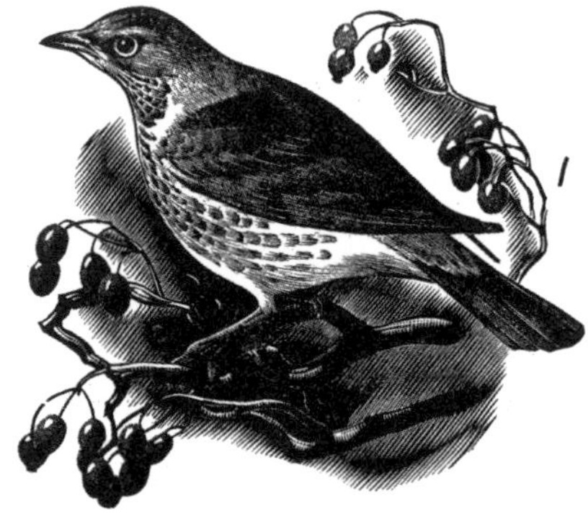

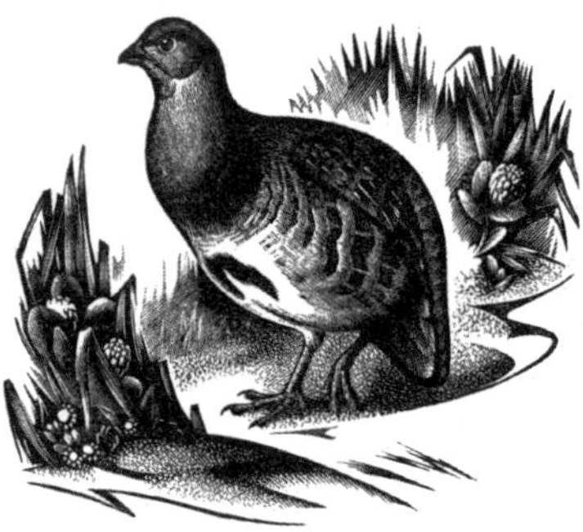

FIG.43a–c
Agnes Miller Parker (clockwise, from top left): *Kite, Fieldfare, Partridge*, 1941. All: wood-engraving on paper. Illustration to Leonard Russell, ed., *The Saturday Book 1941–2: A New Miscellany*, London: Hutchinson, 1941.

contents: 'If anything it looks backward, and nostalgia for small pre-war pleasures emerges from the pages: for days by the sea, for country drives and birds, beasts and flowers… Here, then, is a tranquil book without a thunderclap of any kind. I leave it to the attention of the gentle reader.'[94] Accordingly, the first issue, published in 1941, featured essays by the country writers H.E. Bates and H.J. Massingham, and Gilbert White's descriptions of birds from the Nonesuch Press edition (edited by Massingham). These short texts were accompanied by ten elegant corresponding wood-engravings by Agnes Miller Parker (1895–1980), who had worked with Gertrude Hermes at the Gregynog Press and who had illustrated a number of books concerned with nature and country life over the preceding decade, including H.E. Bates' *Through the Woods* (1936) and *Down the River* (1937), and Thomas Gray's *Elegy Written in a Country Churchyard* (1940). Miller Parker engraved a confident and refined group of birds, reinventing the format of Bewick's vignettes in a style that is entirely her own: an interplay of positive and negative space, between the sinuous lines of the birds' form and the white of the page. Her birds are characterful – we feel not only that we are watching them, but that they are watching us, with a glint of life in their eyes. Miller Parker's birds appear as sentient beings, conscious of our eyes on them and ready to take to the wing. In most cases they are placed within a specific and appropriate natural context. A kite turns to face us whilst perched on a Scots pine (fig.43a); a fieldfare sits on a hawthorn branch surrounded by berries (fig.43b); amidst the heather two grouse peck at the bog; a nightingale rests on an oak branch; and a partridge is framed by clover and grasses, its weight resting on one foot as if the bird is on the threshold of movement (fig.43c).

In March 1941, Penguin Books published an edition of *The Natural History of Selborne* (fig.44) as part of their series of paperbacks, which in 1939 had included Izaak Walton's *Compleat Angler* with illustrations by Gertrude Hermes. Clare Leighton (1898–1989) was commissioned to produce wood-engraved illustrations, and it featured an introduction by the naturalist James Fisher (1912–70), an authority on White who wrote over 40 books on birds and nature during his career.[95] Tacitly noting that White's book represented an escape from the ongoing war, Fisher observed how its appeal was, 'not in what he says so much as in the way he says it. His world is round and simple and complete; the British country; the perfect escape. No breath of the outside world enters in; no politics; no ambition; no care or cost'.[96] Amongst her pictures of nesting birds and harvest mice, or landscapes with farmhouses lashed by rain (fig.118), Leighton stands out for her depictions of women hop picking and working in the fields (fig.45) – subjects mentioned only in passing by White, but featuring in Grimm's engraving *The North East View of Selborne from the Short Lythe* (fig.14).[97] These images of

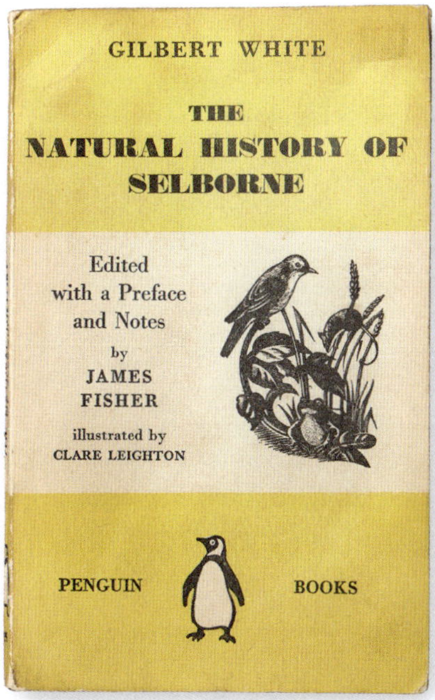

FIG.44
Front cover to Gilbert White, *The Natural History of Selborne Illustrated by Clare Leighton*, James Fisher, ed., Harmondsworth: Penguin, 1941.

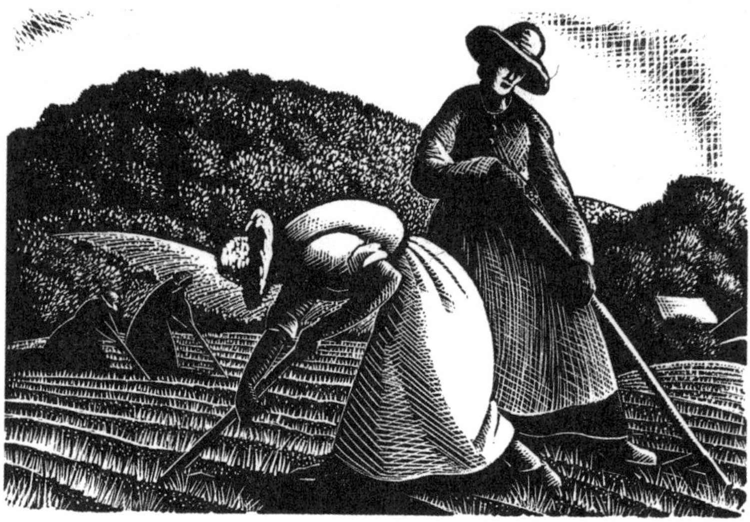

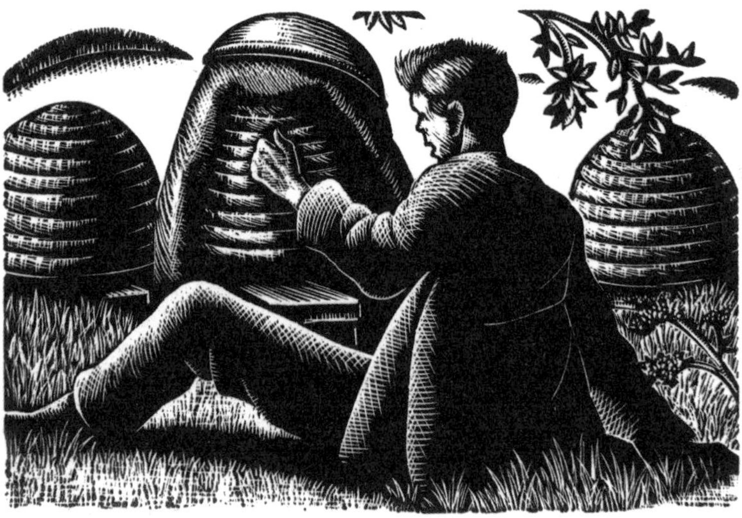

FIG.45 (above)
Clare Leighton, *Women Hoeing,* 1941. Wood-engraving on paper. Illustration to Gilbert White, *The Natural History of Selborne Illustrated by Clare Leighton*, James Fisher, ed., Harmondsworth: Penguin, 1941.

FIG.46 (below)
Clare Leighton, *Boy with Bees,* 1941. Wood-engraving on paper. Illustration to Letter XXVII to the Hon. Daines Barrington, 12 December 1775, in Gilbert White, *The Natural History of Selborne Illustrated by Clare Leighton*, James Fisher, ed., Harmondsworth: Penguin, 1941.

female toil had a timely association with the Women's Land Army (WLA), which made a significant contribution to boosting Britain's food production during the Second World War. With Miller Parker, Leighton had been a key figure in the 1930s revival of wood-engraving, and had written and illustrated several books on country life and rural labour, including *The Farmer's Year* (1933), *Four Hedges* (1935) and *Country Matters* (1937). Leighton emigrated to the USA in 1939 and so worked on her images for the book whilst living on the other side of the Atlantic; comparable images of working women in the American South appear in her book *Southern Harvest (*1942) which she produced soon afterwards. Her interest in the human interaction with nature is reflected in her choice of subjects such as the 'idiot boy' who 'showed a strong propensity to bees' shown with the beehives (fig.46).[98] Unfortunately, the subtlety of the wood-engraved medium was lost in reproduction as the edition was printed on poor quality war-issue paper, and can only be appreciated in the rare, editioned impressions from the woodblocks.

Clare Leighton was an important influence on another female wood-engraver, Claire Oldham (1906–96), a little-known artist who illustrated an edition based on the 1941 Penguin edition by James Fisher which was published by the Cresset Press in April 1947.[99] She illustrated a number of books in the 1940s, but information about her life and work is sparse. Self-taught as an artist, Oldham was the niece of the detective novelist Sir Arthur Conan Doyle. Living near the South Downs, she was working as a land girl during the Second World War when she was apparently 'discovered' by the painter Gluck (1895–1978). Reading Leighton's book on wood-engraving had filled her with 'a desire to hack pictures out of wood' and, according to an article written by the painter Bertram Nicholls that was published in *The Studio* in 1944, 'she had had no training, and had very little leisure, but she had read a book on the subject and did these wood-cuts in what spare time she had.'[100] Nicholls described her wood-engravings as 'a source of admiration and delight', noting that 'I am authoritatively informed, it may be of interest to add, that the birds that figure so effectively in Claire Oldham's work are no mere generalizations, but accurate delineations, faithful, as it were, to a feather, true to the letter as they are certainly true to the spirit'.[101]

Whilst by no means as technically accomplished as Hermes, Leighton or Miller Parker, Clare Oldham's illustrations for *Selborne*

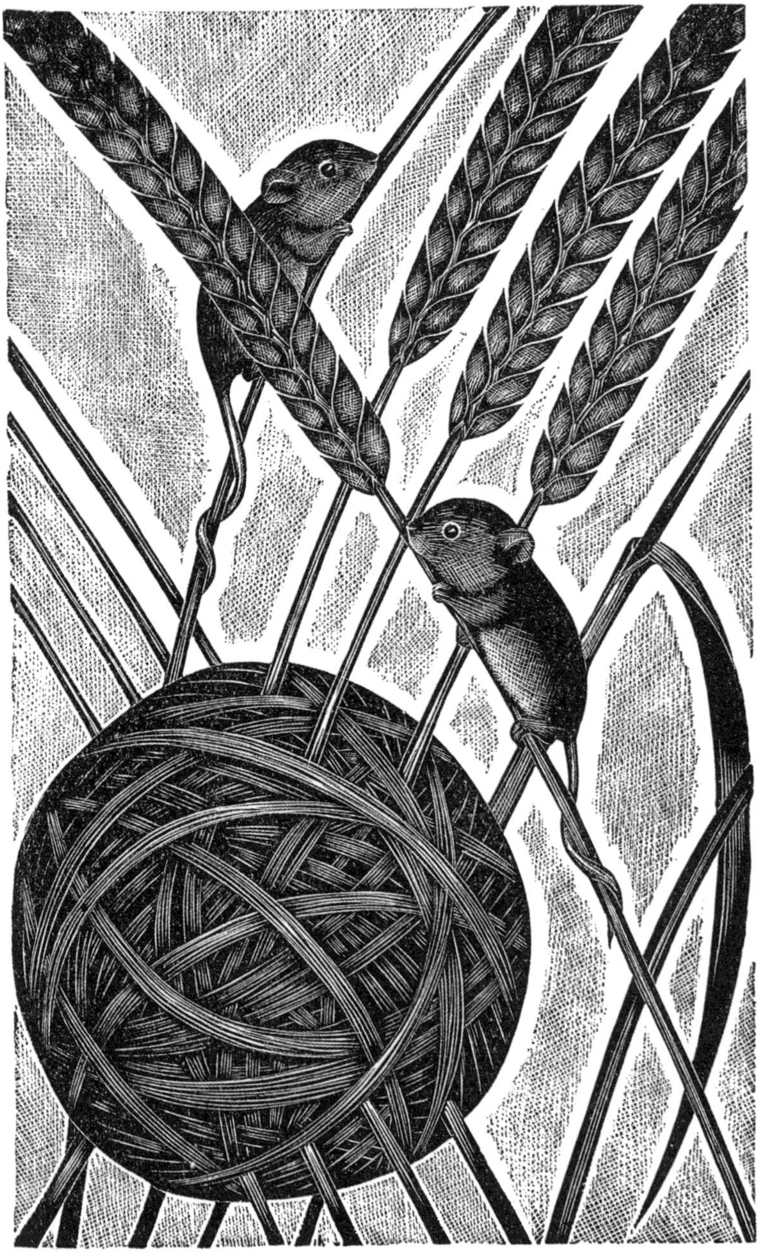

FIG.47
Claire Oldham, *Harvest Mouse and Nest*, 1947. Wood-engraving on paper. Illustration to Letter XII to Thomas Pennant, Selborne, 22 January 1768, in Gilbert White, *The Natural History of Selborne Illustrated by Claire Oldham*, James Fisher, ed., London: The Cresset Press, 1947.

have a particular charm. They include full-page plates featuring sweeping curved arcs as compositional devices, such as a swooping swallow in the opening plate *Road Down to the Village* (fig.70); a fox watching the upward flight of a duck (fig.125); and an array of hedgehogs, including one curled up into a ball, their spines conveyed with especially pleasing mark-making (fig.1).

Although several artists had inevitably chosen to depict harvest mice, Oldham's image (fig.47) is by far the most striking as it so effectively conveys their diminutive scale, showing them climbing ears of wheat above a nest that resembles a tightly wound ball of string, which, in the words of White, was 'most artfully platted, and composed of the blades of wheat, perfectly round and about the size of a cricket-ball'.[102] Similarly, an image of a goldcrest (or 'golden-crowned wren') perched on a yew branch captures White's description of it hanging 'like a titmouse, with its back downward' (fig.48).[103] Oldham conveys 'the wonderful spirit of sociality in the brute creation' described by White in a plate depicting the mismatched friendship between a solitary horse and a hen, who 'spent much of their time together in a lonely orchard, where they saw no creature but each other. By degrees an apparent regard began to take place between these two sequestered individuals' (fig.49).[104] In another equally incongruous pairing, Oldham depicts a

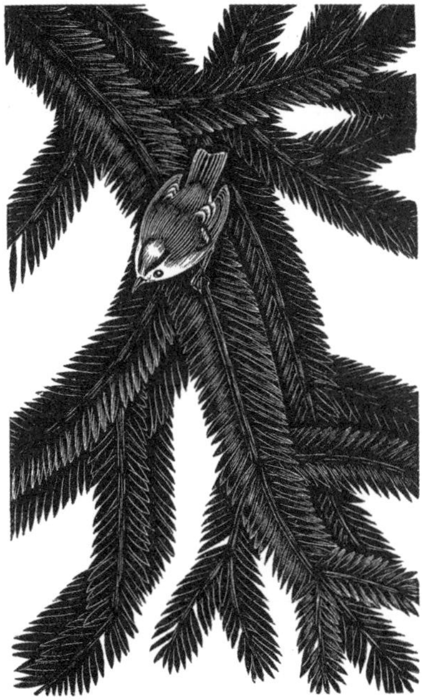

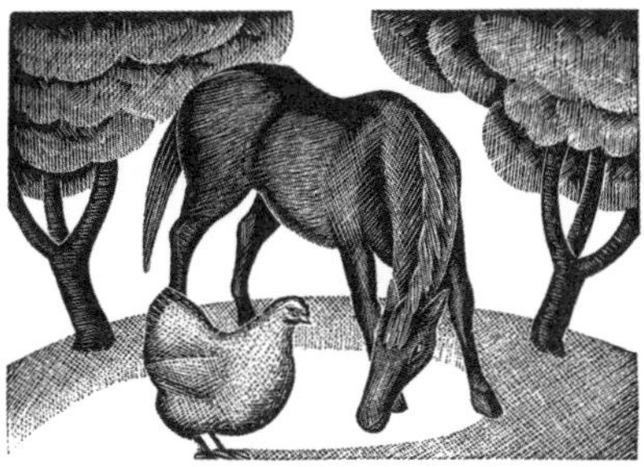

FIG.48
Claire Oldham, *Goldcrest*, 1947. Wood-engraving on paper. Illustration to Letter II to the Hon. Daines Barrington, Selborne, 2 November 1769, Gilbert White, *The Natural History of Selborne Illustrated by Claire Oldham*, James Fisher, ed., London: The Cresset Press, 1947.

FIG.49
Claire Oldham, *Horse and Hen*, 1947. Wood-engraving on paper. Illustration to Letter XXIV to the Hon. Daines Barrington, Selborne, 15 August 1775 in Gilbert White, *The Natural History of Selborne Illustrated by Claire Oldham*, James Fisher, ed., London: The Cresset Press, 1947.

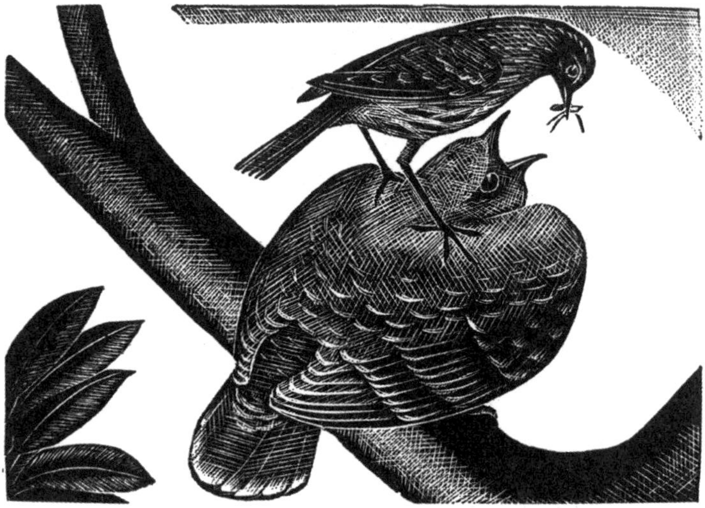

FIG.50
Claire Oldham, *Cuckoo and Meadow-pipit*, 1947. Wood-engraving on paper. Illustration to Letter VII to the Hon. Daines Barrington, in Gilbert White, *The Natural History of Selborne Illustrated by Claire Oldham*, James Fisher, ed., London: Cresset Press, 1947.

diminutive meadow-pipit feeding a comparatively huge cuckoo chick to illustrate White's discovery of a young cuckoo which had been hatched in the nest of a titlark, who was raising it as its own despite it having grown 'vastly too big for its nest' (fig.50).[105]

As a literary classic, Gilbert White's *Natural History* was selected for a 1947 book on typefaces with excerpts from notable texts, each illustrated by a leading contemporary artist. Compiled by the designer and typographer John Lewis (1912–96) for the British printers W.S. Cowell Ltd, it featured illustrations within different typographic settings as an example of what could be achieved through various printing techniques. Contributing artists included Edward Bawden, Barnett Freedman, Blair Hughes-Stanton, Lynton Lamb, Henry Moore, John Nash, John Piper and Graham Sutherland.[106] As Lewis recalled, 'In most cases these passages were decorated with English artists whose work was more often seen in galleries than in books. One of the happiest of these marriages of painter and author was a drawing of John Nash illustrating the letter about the tortoise in Gilbert White's *Selborne*'.[107] Nash (1893–1977) had developed a reputation as a countryman artist due to his illustrations of rural scenes in publications such as William Cobbett's *Rural Rides* (1930), Robert Gathorne-Hardy's *Wild Flowers in Britain* (1938), Adrian

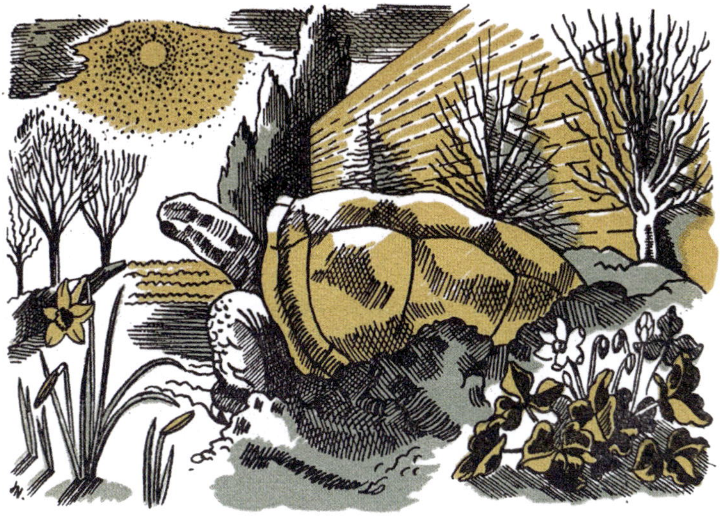

Bell's *Men and the Fields* (1939), and Jason Hill's *The Contemplative Gardener* (1940). His drawing of the tortoise emerging from the earth was printed by letterpress in three colours (fig.51). It inspired the Lutterworth Press to ask if he would undertake an illustrated edition of the complete book, which was to include this image, together with 18 other black and white drawings in the text, 11 full-page drawings, and an imaginary portrait of White as the frontispiece, which was also used on the dustjacket (fig.52).

Nash was known for his botanical illustrations of the plants in his garden, and his 1951 edition of the *Natural History* features pen and ink drawings of flowers such as *Stinking Hellebore* (fig.122) and *Autumn Crocus* (fig.155), as well as topographical drawings of the countryside including views of the woodlands in *Cornfield below the Hanger* (fig.4), *Selborne Church from the Lithe* (fig.140), and *The Stream in the Lithe* (fig.142), amongst other subjects.

FIG.51
John Nash R.A., *The Tortoise, 1947*. Illustration to John Lewis, *A Handbook of Printing Types with Notes on the Style of Composition and Graphic Processes Used by Cowells*, London: Faber & Faber, 1947.

FIG.52
John Nash R.A., cover design for Gilbert White, *The Natural History of Selborne with Drawings by John Nash*, John Lewis, ed., London: Lutterworth Press, London, 1951.

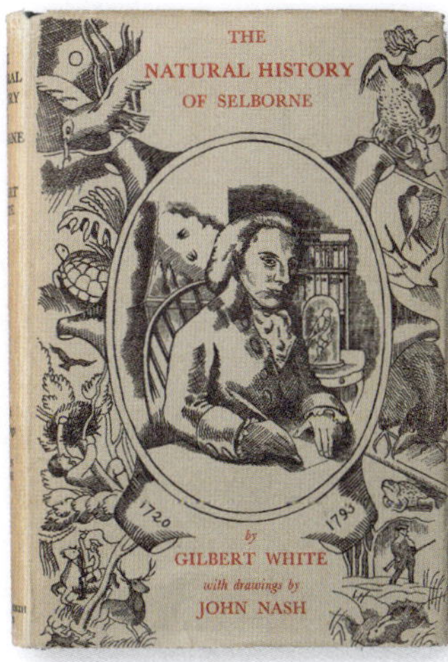

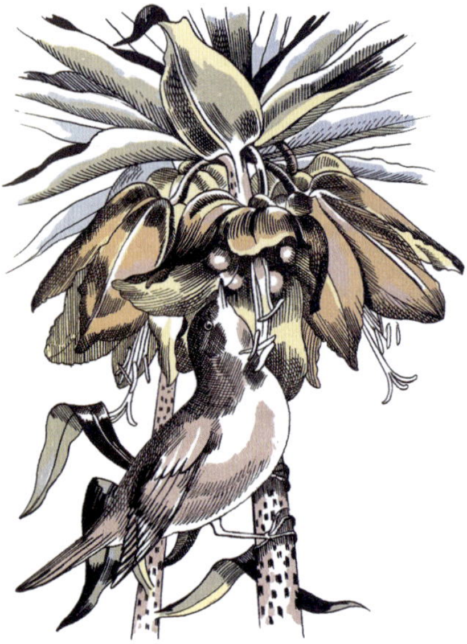

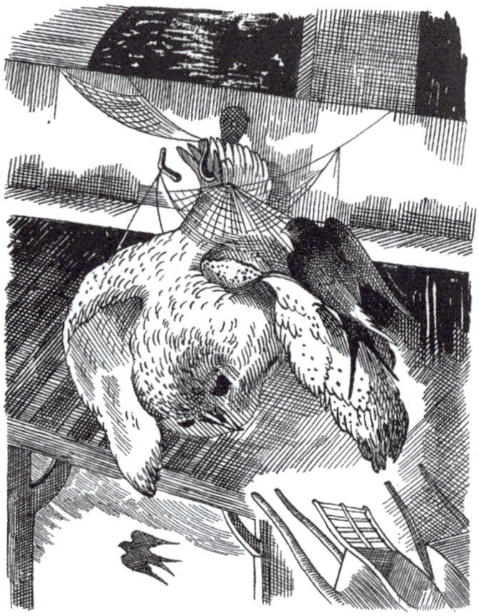

FIG.53
John Nash R.A., '*It runs up the stems of the crown-imperials'*, 1951/72. Lithograph. Illustration to Letter LVII to the Hon. Daines Barrinton, in Gilbert White, *The Natural History of Selborne*, Ipswich: Limited Editions Club, 1972.

FIG.54
John Nash R.A., '*Built its nest on the wings and body of an owl'*, 1951. Illustration to Letter XVIII to the Hon. Daines Barrington, 29 January 1774, in Gilbert White, *The Natural History of Selborne with Drawings by John Nash,* John Lewis, ed., London: Lutterworth Press, London, 1951.

As Clare Colvin has observed, 'Nash's illustrations for this book are distinguished simply because he has penetrated the spirit of White's text and demonstrated his total affinity for the place and situations White was documenting'.[108] Nash depicted scenes that allowed him to integrate his skills as a botanical illustrator, such as the pettichaps that 'runs up the stems of the crown-imperials, and, putting its head into the bells of these flowers, sips the liquor which stands in the nectarium of each petal' (fig.53).[109] He also presented a surreal scene described by White as 'a curiosity worthy of the most elegant private museum in Great Britain' through a densely worked network of cross-hatched lines – the macabre sight of a swallow nesting 'on the wings and body of an owl that happened to hang dead and dry from the rafter of a barn (fig.54)'.[110] In May 1950, Nash wrote to the botanist Clarence Elliott that he was 'nearly blind with cross-hatching in fact I nearly hatched myself off and have now finished 24 pen drawings'.[111]

Humorous incident was at the heart of Nash's images. In one illustration (fig.55), we see a subject previously tackled by

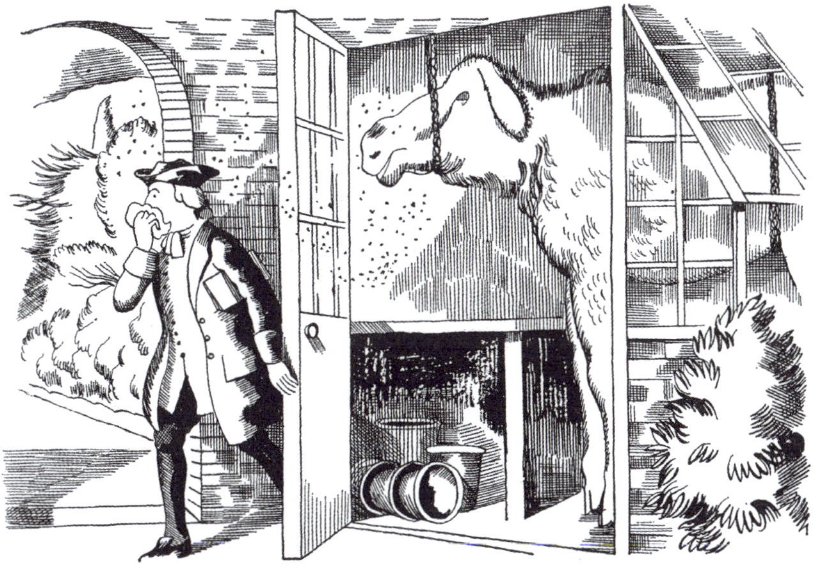

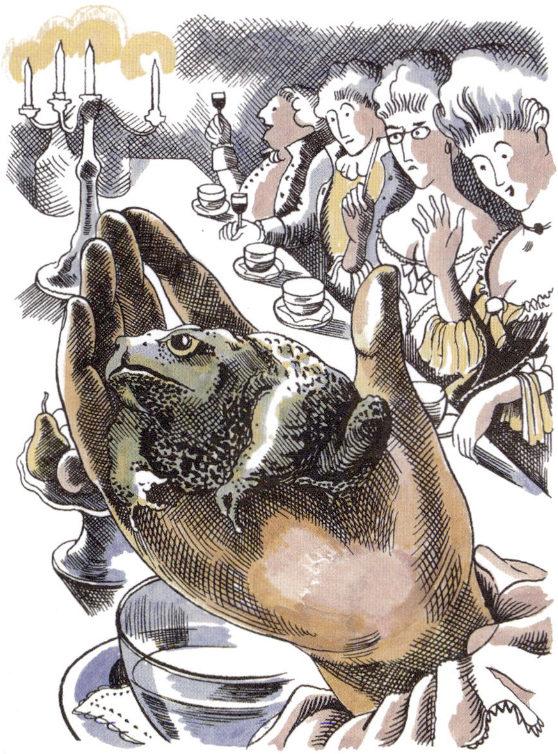

FIG.55
John Nash R.A., '*The dead moose',* 1951. Illustration to Letter XXVIII to Thomas Pennant, March 1770, in Gilbert White, *The Natural History of Selborne with Drawings by John Nash,* John Lewis, ed., London: Lutterworth Press, London, 1951.

FIG.56
John Nash R.A., '*Some ladies.... took a fancy to a toad',* 1951/72. Lithograph on paper. Illustration to Letter XVII, 18 June 1768, in Gilbert White, *The Natural History of Selborne*, Ipswich: Limited Editions Club, 1972.

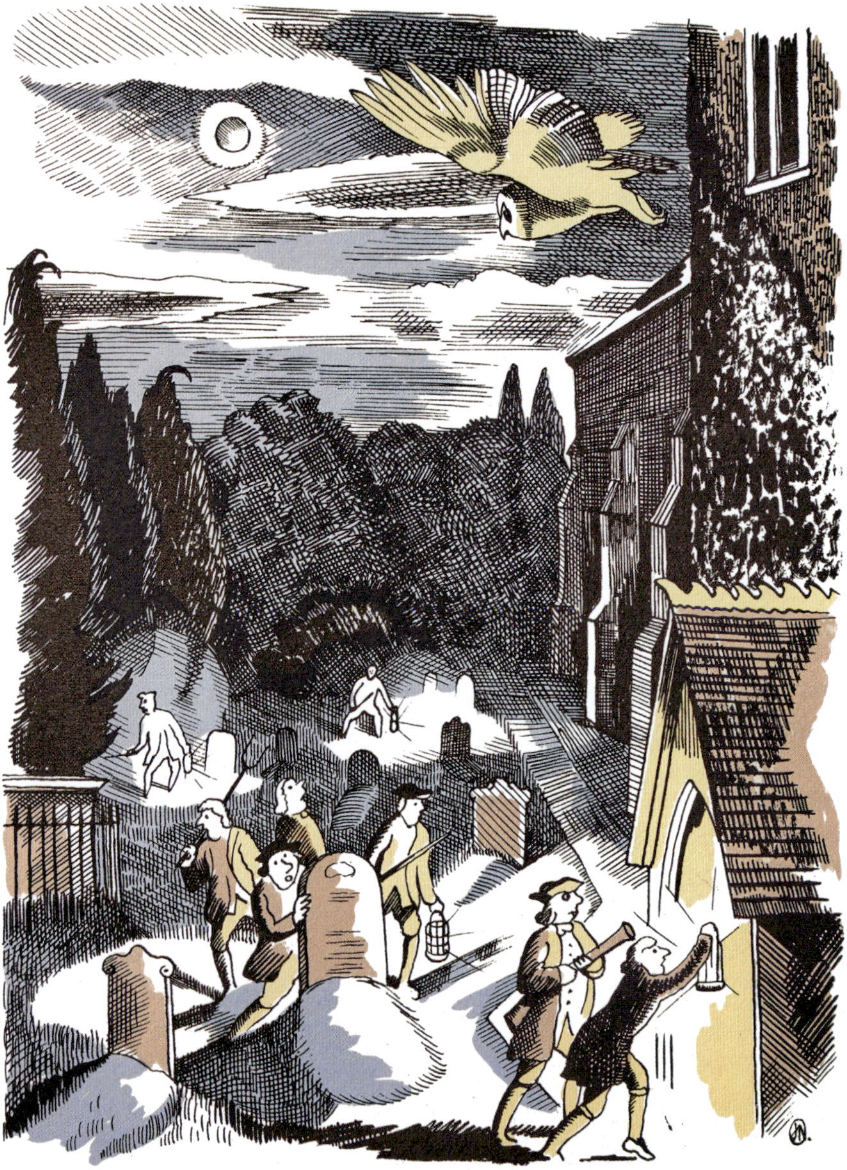

FIG.57
John Nash R.A, *'A whole village up in arms',* 1951/72. Lithograph on paper. Illustration to Letter XV to the Hon. Daines Barrington, 8 July 1773, Gilbert White, *The Natural History of Selborne*, Ipswich: Limited Editions Club, 1972.

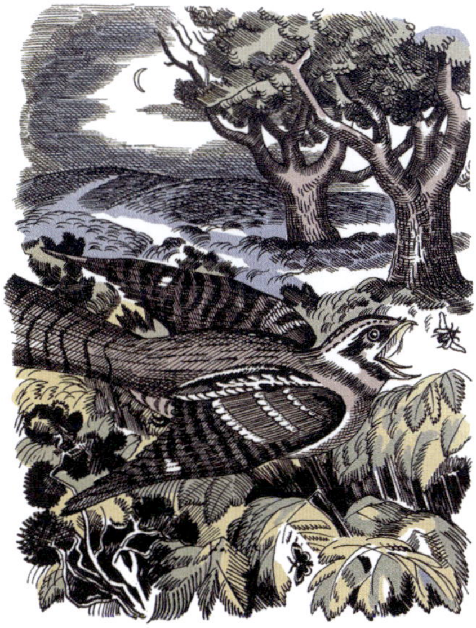

FIG.58
John Nash R.A., '*The powers of its wing were wonderful*', 1951/72. Lithograph on paper. Illustration to Letter XXXVII to Thomas Pennant, 1771, in Gilbert White, *The Natural History of Selborne*, Ipswich: Limited Editions Club, 1972.

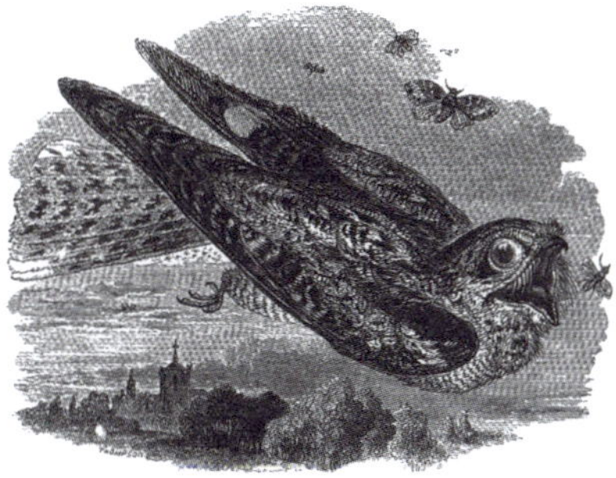

FIG.59
John Thompson, *The Fern-owl*, 1833. Wood-engraving on paper. Illustration to the Lady Drover edition, 1833.

Ravilious in quite a different manner – the naturalist visiting the Duke of Richmond's moose. Nash's version is an example of 'perfect timing' according to Clare Colvin: 'The simple gesture of raising a handkerchief to the nose and the smug grin of the moose convey so exquisitely the ghastly stench emanating from behind the door.'[114] In another memorable illustration, a curious gathering of ladies 'of peculiar taste' who had taken 'a fancy to a toad' look on as the toad is brought to the dinner table for its nightly feeding (fig.56).[113] Elsewhere we see Timothy 'hobbling' towards his benefactress Aunt Rebecca Snookes 'with awkward alacrity' (fig.143).[112] Later we see an aerial view of the churchyard with the 'whole village up in arms' (fig.57) as they search in vain for goblins and spectres after hearing a white owl's 'clamorous hooting'.[115] Lewis observed:

> Painters with a purity and simplicity of vision such as Nash… must be wedded to authors of like vision and stature. Of all contemporary English artists John Nash's work must be the least

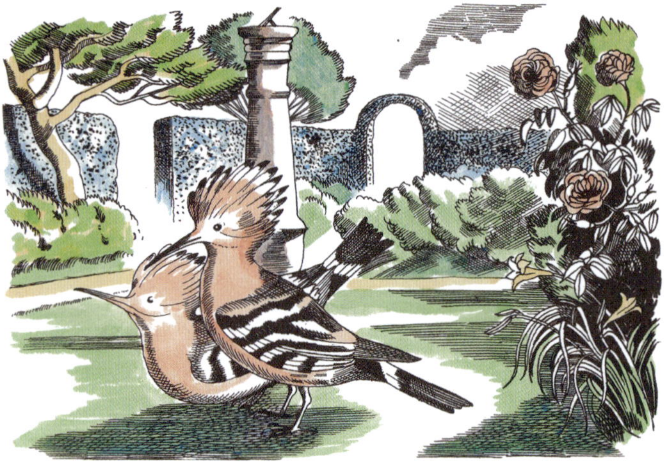

FIG.60
John Nash R.A., '*The most unusual birds…a pair of hoopoes*', 1951/72. Lithograph on paper. Illustration to Letter XI to Thomas Pennant, 9 September 1767, in Gilbert White, *The Natural History of Selborne*, Ipswich: Limited Editions Club, 1972.

FIG.61 (opposite)
Detail and extracts from *The Natural Histories of Selborne by Gilbert White Illustrated with wood engravings by Mrs H.J. Faulkner and W.V. Rose*, Birmingham: City of Birmingham School of Printing, 1950.

affected by the influence of other artists. His vision is a pure distillation of the English scene. I feel that the Rev. Gilbert White would have rejoiced in these beautifully unaffected drawings.[116]

In fact, Nash wasn't averse to quoting other artists. His illustration *The powers of its wing were wonderful* (fig.58) directly quotes a Bewickesque vignette by John Thompson of a flying fern-owl with its beak open to catch a fly (fig.59).

In the 1970s, Nash's illustrations were reissued in a deluxe edition of 1,500 numbered copies, designed by John Lewis and printed by Cowells for the Limited Editions Club.[117] Whereas the earlier edition had been produced on cheap paper in a much smaller format than the original drawings, this edition was unconstrained in this respect, with much larger images so that all the cross-hatched detail could be properly enjoyed. Colour lithographs reproduced the hand-coloured drawings, which are now in the Gilbert White & the Oates Collections, but they required a different approach. Nash explained to Lewis' wife Griselda: 'The two original colour schemes have been stretched a good deal to comply with the needs and I think the actual colours will have to be modified in the printing. You see we have nothing approaching a Red or a Pink and this lets us down over the damned Hoopoes.'[118] The end result is arguably one of the most

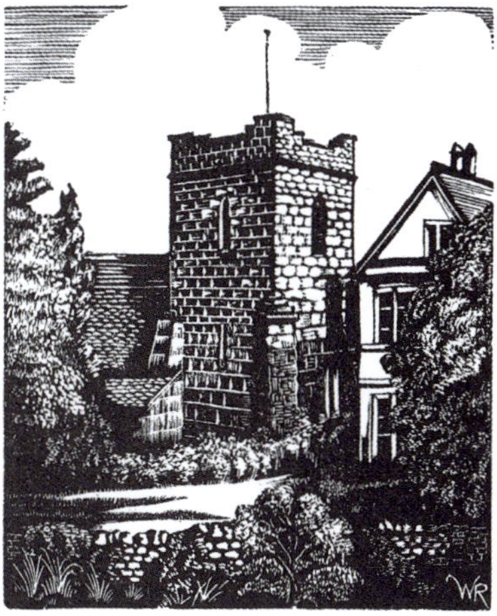

sumptuous, spacious and luxurious of all editions of the *Natural History.*

With hundreds of published editions of the *Natural History* and such numerous artists responding to and illustrating the same text, the question of reproduction and originality can never be far away. From the transposition of Grimm's drawings and Elmer's painting into engravings, to the copying of Bewick's vignettes and homages to past masters, it is clear that providing an original viewpoint that complements White's text is not without its challenges. The most successful have remained complementary and distinct largely because the artists remained true to their own interests and vision. Inevitably over the years, artists have chosen to show the same subjects as those who have gone before, such as the multitude of depictions of Timothy the Tortoise, of barn owls and harvest mice, the boy 'birdsnesting' (collecting eggs), which both Ravilious and Leighton had tackled.[119]

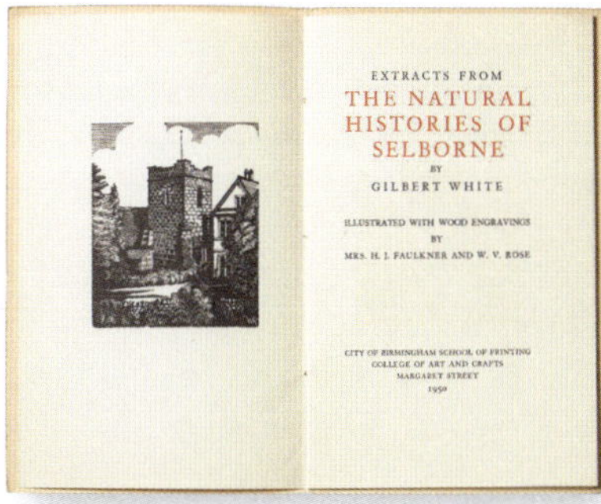

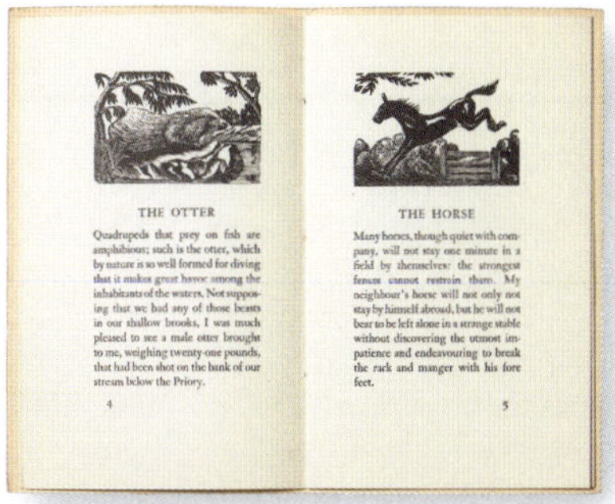

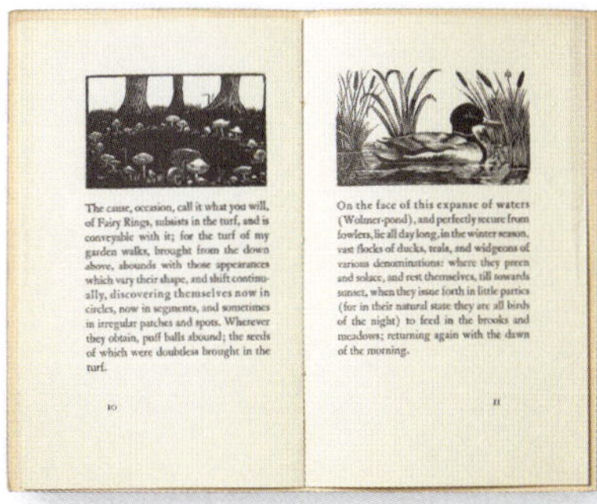

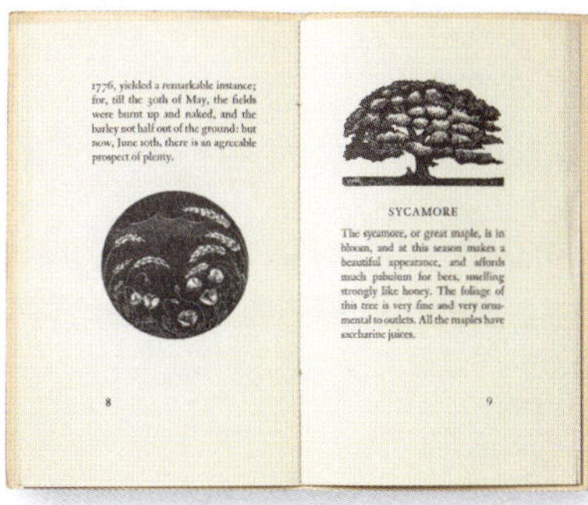

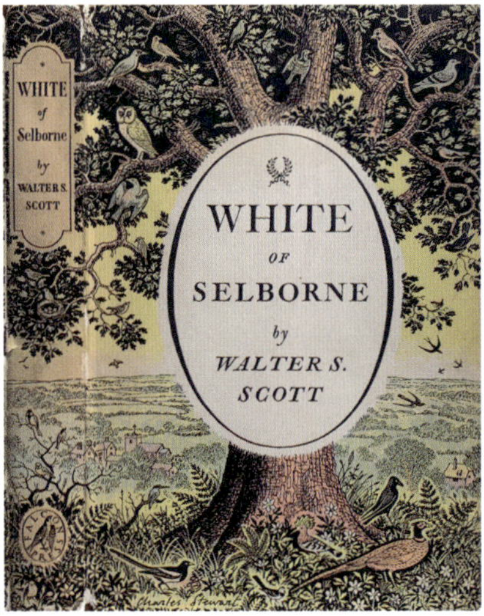

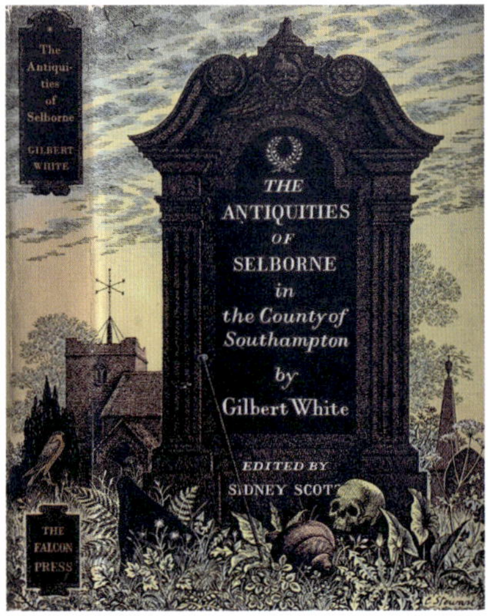

FIGS 62 & 63
Charles Stewart, covers for Gilbert White, *The Antiquities of Selborne in the County of Southampton*, Sidney Scott, ed., London: The Falcon Press, 1950, and Walter S. Scott, *White of Selborne*, London: The Falcon Press, 1950.

The book has also provided a stimulating exercise for art students, as with a booklet produced by staff and students of the Birmingham School of Printing in 1950 where excerpts from White's *Natural History* were used as inspiration for their somewhat crude wood-engravings (fig.61).[120] In the same year, the co-principal of the Byam Shaw School of Art, Charles Stewart (1915–2001), produced two cover designs for Falcon Press – an edition of *The Antiquities* featuring a macabre scene in Selborne churchyard with a discarded tricorn hat, walking stick and skull propped against a gravestone bearing the book's title (fig.62), and Walter S. Scott's book

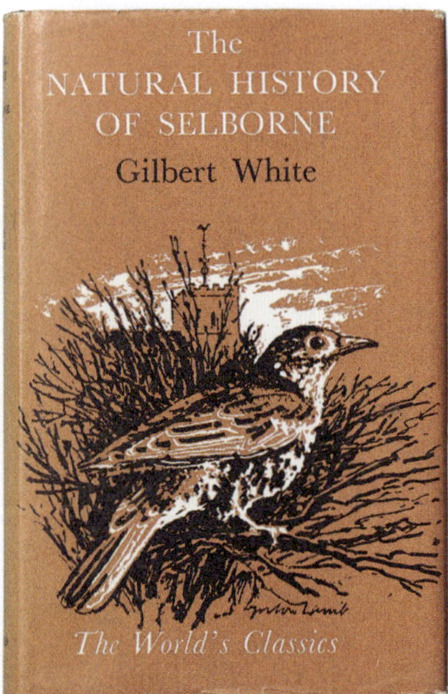

FIG.64
Lynton Lamb RDI, *White's Thrush*. Cover design for The World's Classics edition of Gilbert White, *The Natural History of Selborne*, Oxford: Oxford University Press, 1977.

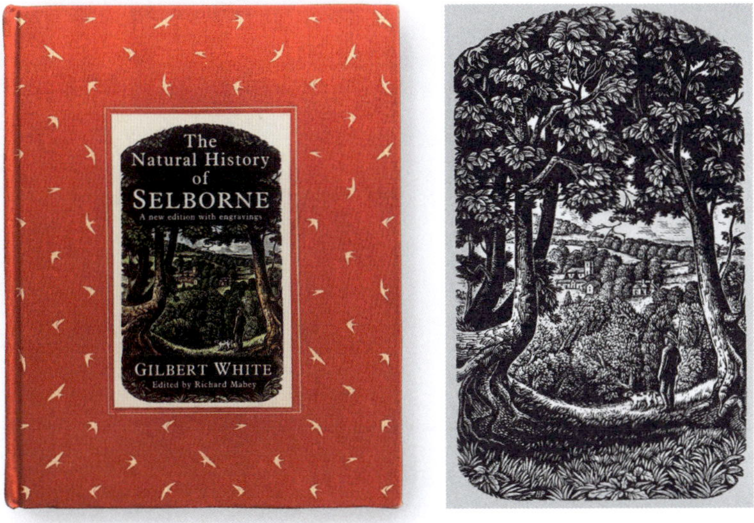

FIG.65 AND DETAIL
Howard Phipps, *Selborne*, 1988. Wood-engraving on paper, cover design for Gilbert White, *The Natural History of Selborne: A new edition with engravings*, Richard Mabey, ed., London, Melbourne, Auckland, Johannesburg: Century, 1988. Cover design by Fielding Rowinski

White of Selborne, which is adorned with a bird-filled tree before an expansive landscape (fig.63). Lynton Lamb (1907–77), Head of Lithography at the Royal College of Art, illustrated Cecil Emden's 1956 book *Gilbert White in his Village* with imaginary images of the parson at work in his gardens. He also designed a cover for the compact 1977 Oxford University Press edition that featured White's thrush (*Zoothera aurea*), a bird named after the naturalist (fig.64). In the same year, the Gilbert White Museum published an edition featuring reproductions of ten detailed drawings of Selborne by Frederick Marns (1933–2013).[121] The wood engraver Howard Phipps (b.1954) produced a view looking back towards the village from the Hanger (fig.65) for the cover of a 1988 volume edited by Richard Mabey.[122] There have been numerous others editions including international publications, such as in Denmark illustrated by Mads Stage, in Sweden illustrated by Gunnar Brusewitz, and in Japan, where White is widely revered.

For the 1962 Folio Society edition edited by W.S. Scott, John Piper (1903–91) took a singular route in the modern era, with a series of 28 pen, ink and wash drawings that were reproduced in half-tone collotype.[123] These are not detailed images of flora or

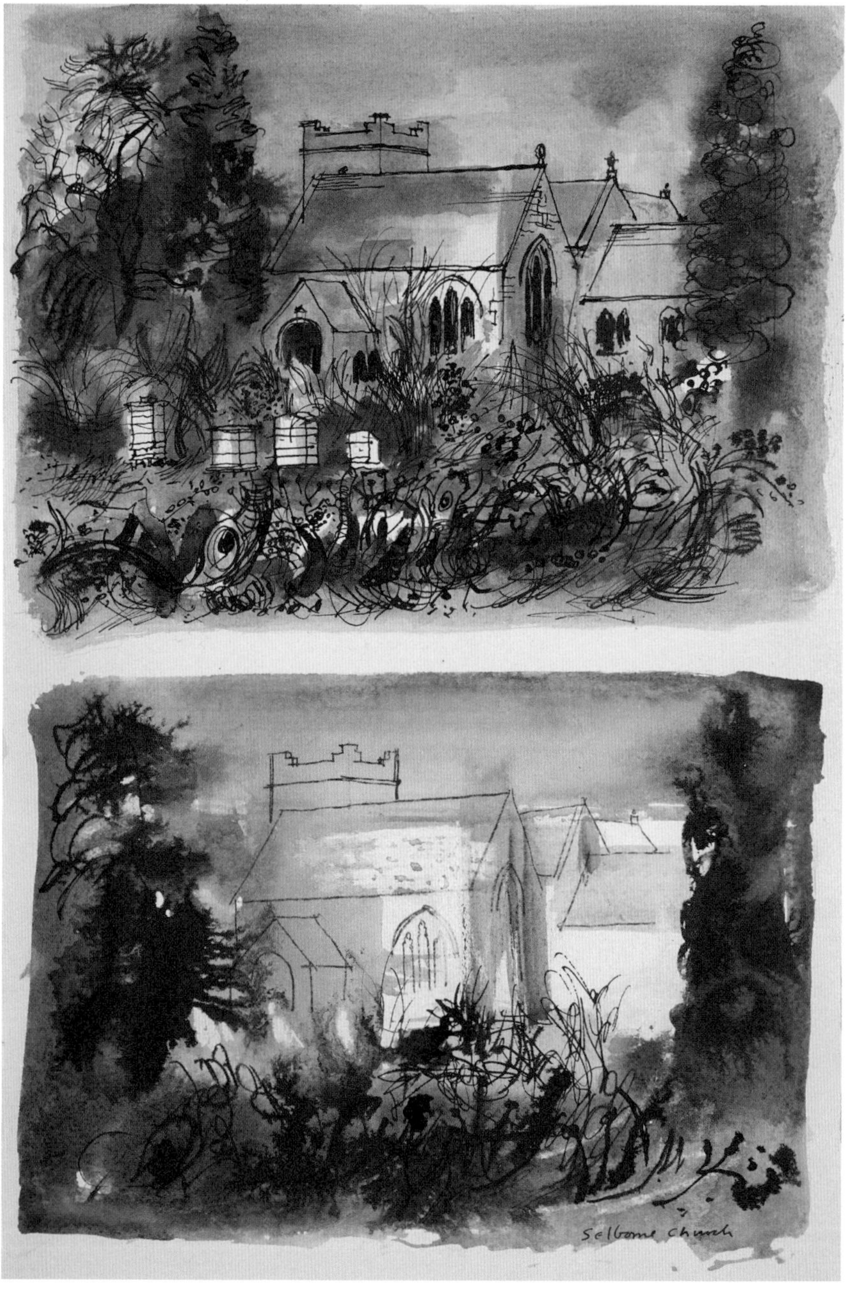

FIG.66

John Piper, *Selborne Church, 1961*. Ink and wash drawing, Gilbert White, *The Natural History of Selborne*, W.S. Scott ed., London; Folio Society, 1961.

fauna; there is none of the particularity that Piper had seen as manifesting the Romantic spirit in Bewick's work, nor scenes of the interaction of man with nature. Instead, Piper produced groups of loosely executed studies of views that he had encountered in the village, often several to a sheet, proposing alternative visual descriptions of the same subject with varying degrees of detail. These include studies of tree roots in the wood at Daton, the barley fields and Hawkley hanger, hopfields, and subjects that reflected his interest in architecture, such as The Wakes and a view of beehives in front of St Mary's church (fig.66). In the words of Joanna Selborne, they are 'perhaps the most idiosyncratic and abstract of all. His pen and wash drawings... though attractive are barely evocative of Selborne'.[124]

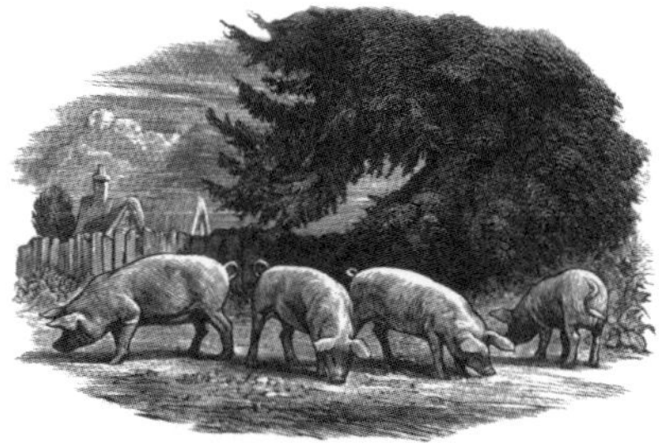

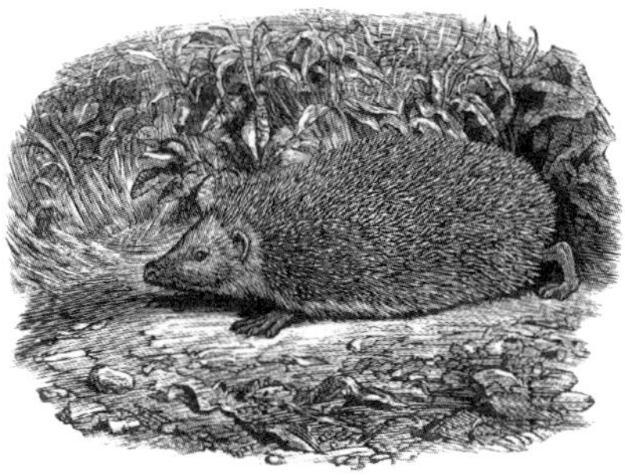

FIGS 67 & 68
Christopher Wormell, *Pigs and Yew Trees* (above), *Hedgehog* (below), 1994. Both: wood-engraving. Illustration to Gilbert White, *The Natural History of Selborne*, Ian Niall, ed. London: The Folio Society, 1994.

A subsequent Folio Society edition, introduced by Ian Niall in 1994, took almost the opposite approach with a series of highly detailed and naturalistic wood-engravings by the self-taught printmaker Christopher Wormell (b.1955), who had previously worked with Niall on *English Country Traditions* (1988) and had illustrated other countryside books, such as a 1986 edition of Henry Williamson's *Story of a Norfolk Farm*. Wormell's approach provides an arc back to Thomas Bewick's vignettes, with small and detailed engravings of livestock, including a sow (fig.164), cockerel (fig.165) and horse (fig.166), and more specific creatures described by White such as the stone curlews at Chichester (fig.168) and a tortoise (fig.169). Wormell focussed in on the particular in White's writings, and rendered it in exquisite detail.

Contemporaneous with these engravings, in 1993 a memorial window was installed in the Lady Chapel of St Mary's Church in Selborne to mark the bicentenary of White's death. It was described as being 'among the loveliest of 20th-century windows' by Simon Jenkins.[125] The clear glass etched with images of leaves allows a

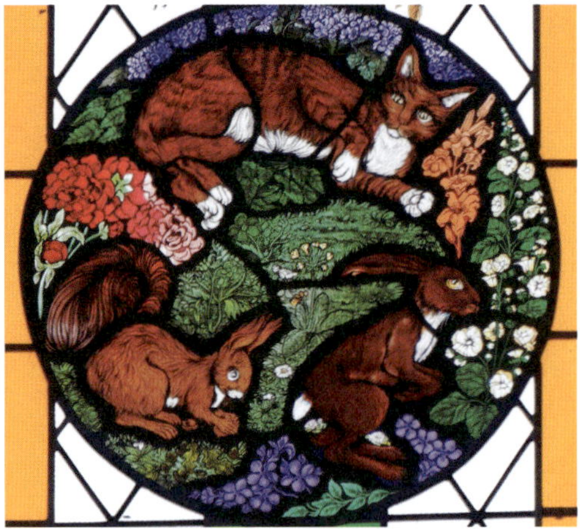
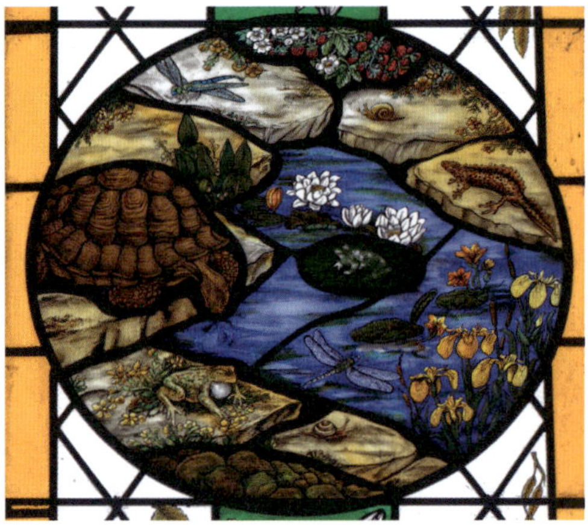
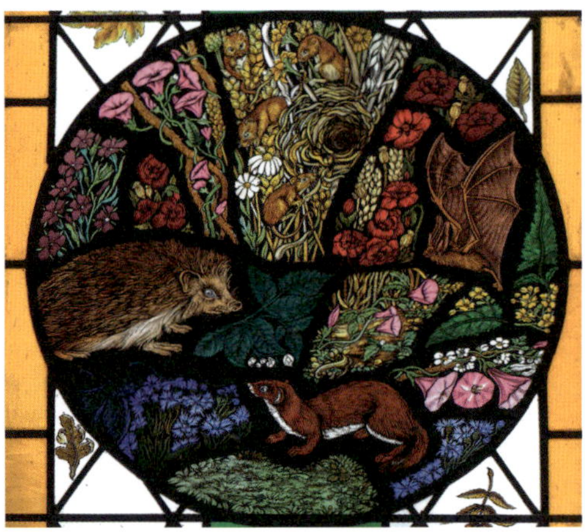

FIG.69a–c
Roundels by Harold Thompson, John Tarrant and Jude Tarrant, *Cat, Hare and Squirrel* (top); *Tortoise and Pond Life* (middle); *Harvest Mice, Hedgehog, Stoat and Noctule Bat* (bottom), Gilbert White Bicentenary memorial window. All: etched and painted stained glass, Lady Chapel, St Mary's Church, Selborne.

view of nature beyond, with three colourful stained-glass roundels conceived by Harold Thomson and executed by John and Jude Tarrant. One depicts the wild mammals in the fields: a family of harvest mice building their nest, a stoat, a hedgehog and a long-eared bat (fig.69a); another the reptiles and insects that White found interesting, including his own tortoise Timothy (eventually found to be female) and the Hampshire Natterjack toad (fig.69b); and the third, the naturalist's story of a family of young squirrels nurtured by a cat (fig.69c). At the foot is a panel inscribed:

> *Gilbert White 1720–1793*
> *God be praised for a faithful priest*
> *A humble student of nature*
> *And a writer of genius.*

In the Advertisement to the first edition Gilbert White wrote:

> If the writer should at all appear to have induced any of his readers to pay a more ready attention to the wonders of the Creation, too frequently overlooked as common occurrences; or if he should by any means, through his researches, have lent a helping hand towards the enlargement of the boundaries of historical or topographical knowledge… his purpose will be fully answered.[126]

Three hundred years on White's ambition has not only been fulfilled, but he has achieved far more than he might have ever imagined. It is testament to the elusive country parson–naturalist that after 400 editions and reissues of his *Natural History,* his book is still able to engage new readers. Generation after generation has been drawn to nature by White's writings. So many artists have been inspired by his perfectly described observations of the natural life in the village of Selborne and its surrounding area, and each has reflected the art of their times, so that the various editions of his book form their own microcosm of British natural history illustration. With species extinction and loss of biodiversity, global warming, and environmental catastrophe facing the natural world, the message of 'the first ecologist' seems to be more appropriate than ever. We should notice and take care of the nature on our doorsteps, before it is lost.

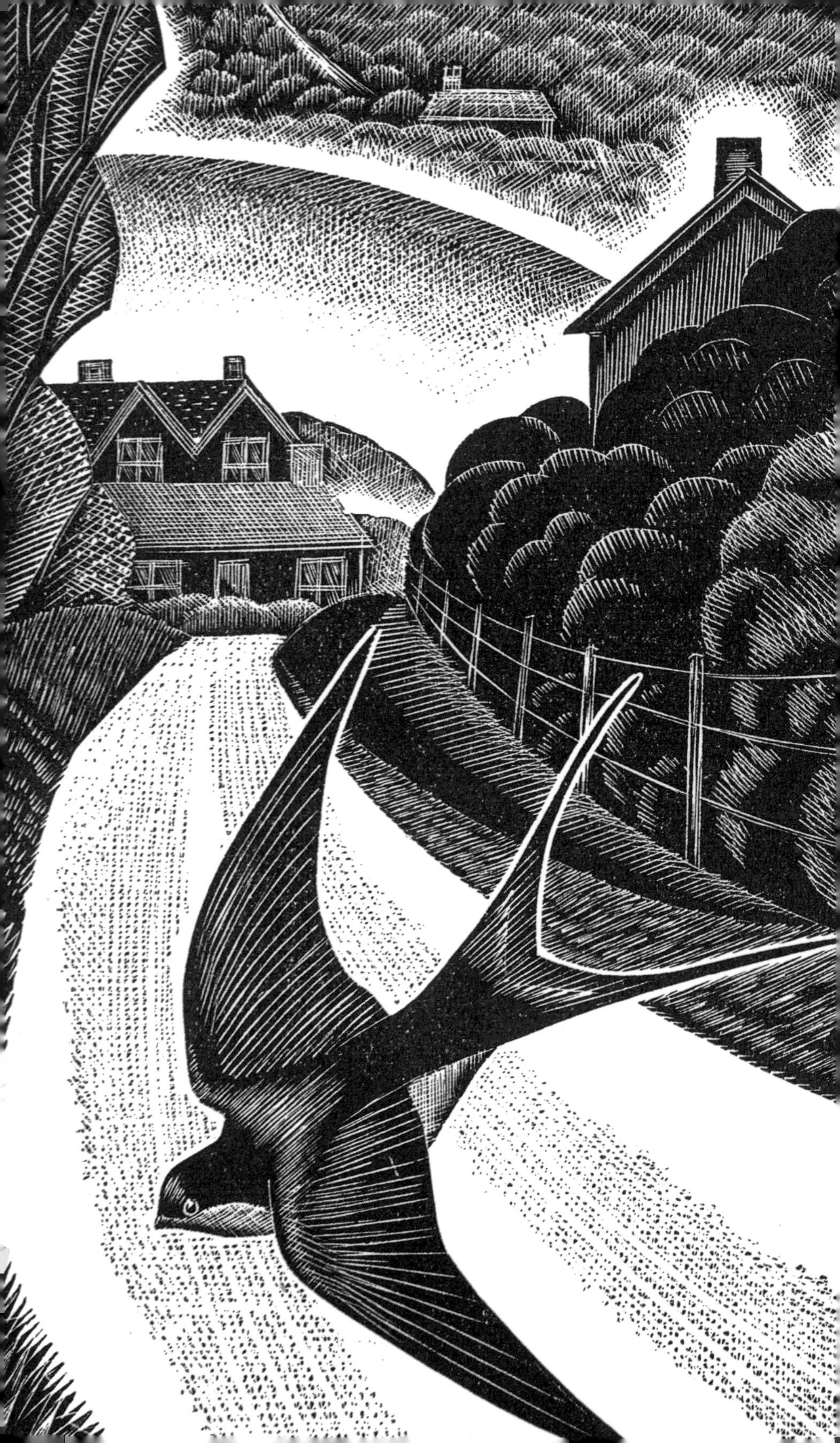

3.
WHITE'S SELBORNE

Virginia Woolf (1882–1941)

First published in *New Statesman and Nation*, 30 September 1939

There is somewhat in most genera at least, that at first sight discriminates them, and enables a judicious observer to pronounce upon them with some certainty.' Gilbert White is talking, of course, about birds; the good ornithologist, he says, should be able to distinguish them by their air – 'on the ground as well as on the wing, and in the bush as well as in the hand.' But when the bird happens to be Gilbert White himself, when we try to discriminate the colour and shape of this very rare fowl, we are at a loss. Is he, like the bird so brightly coloured by hand as a frontispiece to the second volume, a hybrid – something between a hen that clucks and a nightingale that sings? It is one of those ambiguous books that seem to tell a plain story, *The Natural History of Selborne,* and yet by some apparently unconscious device of the author's

FIG.70
Claire Oldham, '*Road down to the Village*', 1947. Wood-engraving, from *The Natural History of Selborne by Gilbert White edited and with an introduction by James Fisher,* London: Cresset Press, 1947.

has a dog barking, cart wheels creaking, and see, when 'all the fading landscape sinks in night', if not Venus herself, at least a phantom owl.

His intention seems plain enough – it was to impart certain observations upon the fauna and flora of his native village to his friends Thomas Pennant and Daines Barrington. But it was not for the benefit of those gentlemen that he composed the sober yet stately description of Selborne with which the book opens. There it is before us, the village of Selborne, lying in the extreme eastern corner of the county of Hampshire, with its hanger and its sheep walks and those deep lanes 'that affright the ladies and make timid horsemen shudder'. The soil is part clay, part malm; the cottages are of stone or brick; the men works in the hop gardens and in the spring and summer the women weed the corn. No novelist could have opened better. Selborne is set solidly in the foreground. But something is lacking; and so before the scene fills with birds, mice, voles, crickets, and the Duke of Richmond's moose, before the page is loud with the chirpings, bleatings, lowings, and gruntings of their familiar intercourse, we have Queen Anne lying on the bank to watch the deer driven past. It was an anecdote, he casually remarks, that he had from an old keeper, Adams, whose great grandfather, father and self were all keepers in the forest. And thus the single straggling street is allied with history, and shaded by tradition. No novelist could have given us more briefly and completely all we need to know before the story begins.

The story of Selborne is a vegetable, an animal story. The gossip is about the habits of vipers and the love interest is supplied chiefly by frogs. Compared with Gilbert White the most realistic of novelists is a rash romantic. The crop of the cuckoo is examined; the viper is dissected; the grasshopper is sought with a pliant grass blade in its hole; the mouse is measured and found to weigh one copper halfpenny. Nothing can exceed the minuteness of these observations, or the scrupulous care with which they are conducted. The chief question in dispute – it is indeed the theme of the book – is the migration of swallows. Barrington believed that the swallow sleeps out the winter; White, who has a nephew in Andalusia to inform him, now inclines to migration; then draws back. Every grain of evidence is sifted; none is obscured. With all his faculties bent on this great question, the image of science at her most innocent and most sincere, he loses that self-consciousness which so often separates us from our fellow creatures and becomes like a bird seen through a field-glass busy in a distant hedge. This is the moment then, when his eyes are fixed upon the swallow, to watch Gilbert White himself. We observe in the first place the creature's charming simplicity. He is quite indifferent to public opinion. He will transplant a colony of crickets to his lawn; imprison one in a paper cage on his table; bawl through a speaking trumpet at his bees – they remain indifferent; and arrive at Selborne with Aunt Snooke's aged tortoise seated beside him in the post chaise. And while thus engaged he emits those chuckles of delight, those half-conscious burblings and comments which make him as 'amusive' as one of his own birds. '.... But their inequality of height,' he muses, pondering the abortive match between the moose and the red deer, 'must always have been a bar to any commerce of an amorous kind.'

> 'The copulation of frogs,' he observes, 'is notorious to everybody... and yet I never saw, or read, of toads being observed in the same situation.'

'Pitiable seems the condition of this poor embarrassed reptile,' he laments over the tortoise, yet there is a season (usually the beginning of June) when the tortoise walks on tip-toe along the garden path in search of love. And just as the vicarage garden seemed to Aunt Snooke's tortoise a whole world, so, as we look through the eyes of Gilbert White, England becomes immense. The South Downs, across which he rides year after year, turn to 'a vast range of mountains'. The country is very empty. He is more solitary at Selborne than a peasant to-day in the remotest Hebrides. It is true that he has – he is proud of the fact – a nephew in Andalusia; but he has no acquaintance among the gentlemen of the Navy; and though London and Bath exist, of course – London indeed boasts a very fine collection of horns – rumours from those capitals come very slowly across wild moors and roads which the snow has made unpassable. In this quiet air sounds are magnified. We hear the whisper of grasshopper lark; the caw of rooks is like a pack of hounds 'in hollow, echoing woods'; and on a still summer evening the Portsmouth gun booms out just as the goat-sucker begins its song. His mind, like the bird's crop that the farmer's wife found stuffed with vegetables and cooked for her dinner, has nothing but insects in it and tender green shoots. This innocent, this unconscious happiness is conveyed, not by assertion, but much more effectively by those unsought memories that come of their own accord. They are all of hot summer evenings – at Oxford in Christ Church quadrangle; riding from Richmond to Sunbury with the swallows skimming the river. Even the strident voice of the cricket, so discordant to some, fills his mind 'with a train of summer ideas, of everything that is rural, verdurous and joyful'. There is a continuity in his happiness; the same thoughts recur on the same occasions. 'I made the same remark in former years as I came the same way annually.' Year after year he was thinking of the swallows. But the landscape in which this bird roams so freely has its hedges. They shut in, but they protect. There is what he calls, so aptly, Providence. Church spires, he remarks, 'are very necessary ingredients in the landscape.' Providence dwells there – inscrutable, for why does it allot so many years to Aunt Snooke's tortoise? But all – wise – consider the legs of the frog – 'How wonderful is the economy of Providence with regards to the limbs of so vile an animal!' In another fifty years Providence would have been neither so inscrutable nor as wise – it would have lost its shade. But Providence about 1760 was in its prime; it sets all doubts at rest, and so leaves the mind free to question practically everything. Besides Providence there are the castles and seats of the nobility. He respects them almost equally. The old families – the Howes, the Mordaunts – know their places and keep the poor in theirs. Gilbert White is far less tender to the poor – 'We abound with poor ,' he writes, as if the vermin were beneath his notice – than to the grasshopper whom he lifts out of its hole so carefully and once inadvertently squeezed to death. Finally, shading the landscape with its august laurel, is literature – Latin literature, naturally. His mind is haunted by the classics. He sounds a Latin phrase now and then as if to tune his English. The echo that was so famous a feature of Selborne seems of its own accord to boom out *Tityre, tu patulae recubans...* It was with Virgil in his mind that Gilbert White described the women making rush candles at Selborne. So we observe through our field-glasses this very fine specimen of the eighteenth-century clerical naturalist. But just as we think to have got him he moves. He sounds a note that is not the characteristic note of the common English clergyman. 'When I hear fine music I am haunted with passages there from night and day: and especially at first waking, which by their importunity, give me more uneasiness than pleasure.' Why does music, he asks, 'so strangely affect some men, as it were by recollection, for days after a concert is over?' It is a question that sends us baffled to his biography. But we learn only what we knew already – that his affection for Kitty Mulso was not passionate; that he was born at Selborne in 1720 and died there in 1793; and that his 'days passed with scarcely any other vicissitudes than those of the seasons.' But one fact is added – a negative, but a revealing fact; there is no portrait of him in existence. He has no face. That is why perhaps he escapes identification. His observation of the insect in the grass is minute; but he also raises his eyes to the horizon and looks and listens. In that moment of abstraction he hears sounds that make him uneasy in the early morning: he escapes from Selborne, from his own age, and comes winging his way to us in the dusk along the hedgerows. A clerical owl? A parson with the wings of a bird? A hybrid? But his own description fits him best. 'The kestrel or windhover,' he says, 'has a peculiar mode of hanging in the air in one place his wings all the time being briskly agitated.'

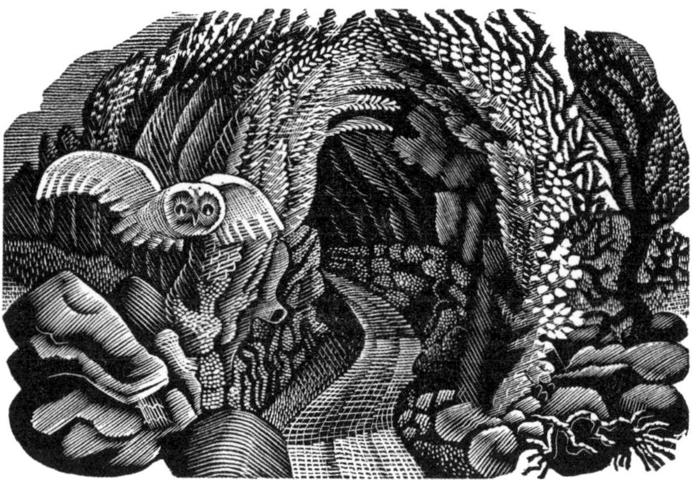

FIG.71
Eric Ravilious, *Hollow Lane,* 1938. Wood-engraving on paper. Illustration to *The Writings of Gilbert White*, H.J. Massingham, ed., London: The Nonesuch Press: 1938.

POSTHUMOUS LETTER TO GILBERT WHITE (1973)

W.H. Auden

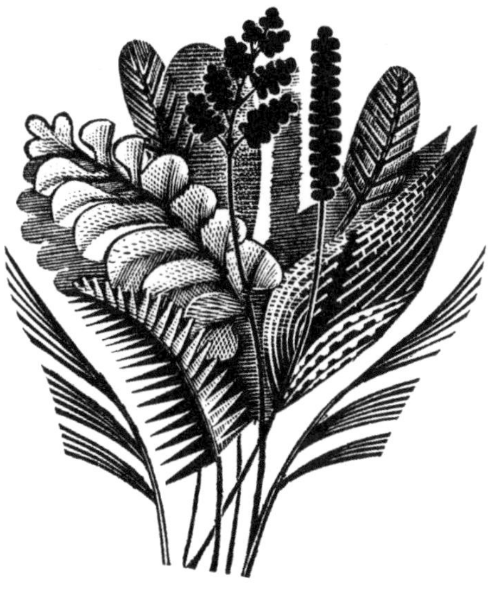

FIG.72
Eric Ravilious, *Tailpiece*, 1938. Wood-engraving on paper, in H.J. Massingham, ed., *The Writings of Gilbert White – Selected and Edited with an Introduction by H.J. Massingham*, London: The Nonesuch Press, 1938.

It's rather sad we can only meet people
whose dates overlap with ours, a real shame that
you and Thoreau (we know that he read you)
never shook hands. He was, we hear, a rabid

Anti-Clerical and quick-tempered, you the
quietest of curates, yet I think he might well have
found in you the Ideal Friend he wrote of
with such gusto, but never ran into.

Stationaries, both of you, but keen walkers,
chaste by nature and, it would seem, immune to
the beck of worldly power, kin spirits,
who found all creatures amusive, even

the tortoise in spite of its joyless stupors,
aspected the vagrant moods of the Weather,
from the modest conduct of fogs to
the coarse belch of thunder or the rainbow's

federal arch, what fun you's have had surveying
two rival landscapes and their migrants, noting
the pitches owls hoot on, comparing
the echo-response of dactyls and spondees.

Selfishly, I, too, would have plumbed to know you:
I could have learned so much. I'm apt to fancy
myself as a lover of Nature,
but have no right to, really. How many

birds and plants can I spot? At most two dozen.
You might, though, have found such an ignoramus
a pesky bore. Time spared you that: I
have, though, thank God, the right to re-read you.

(Published in *The New York Review,* 18 October 1973)

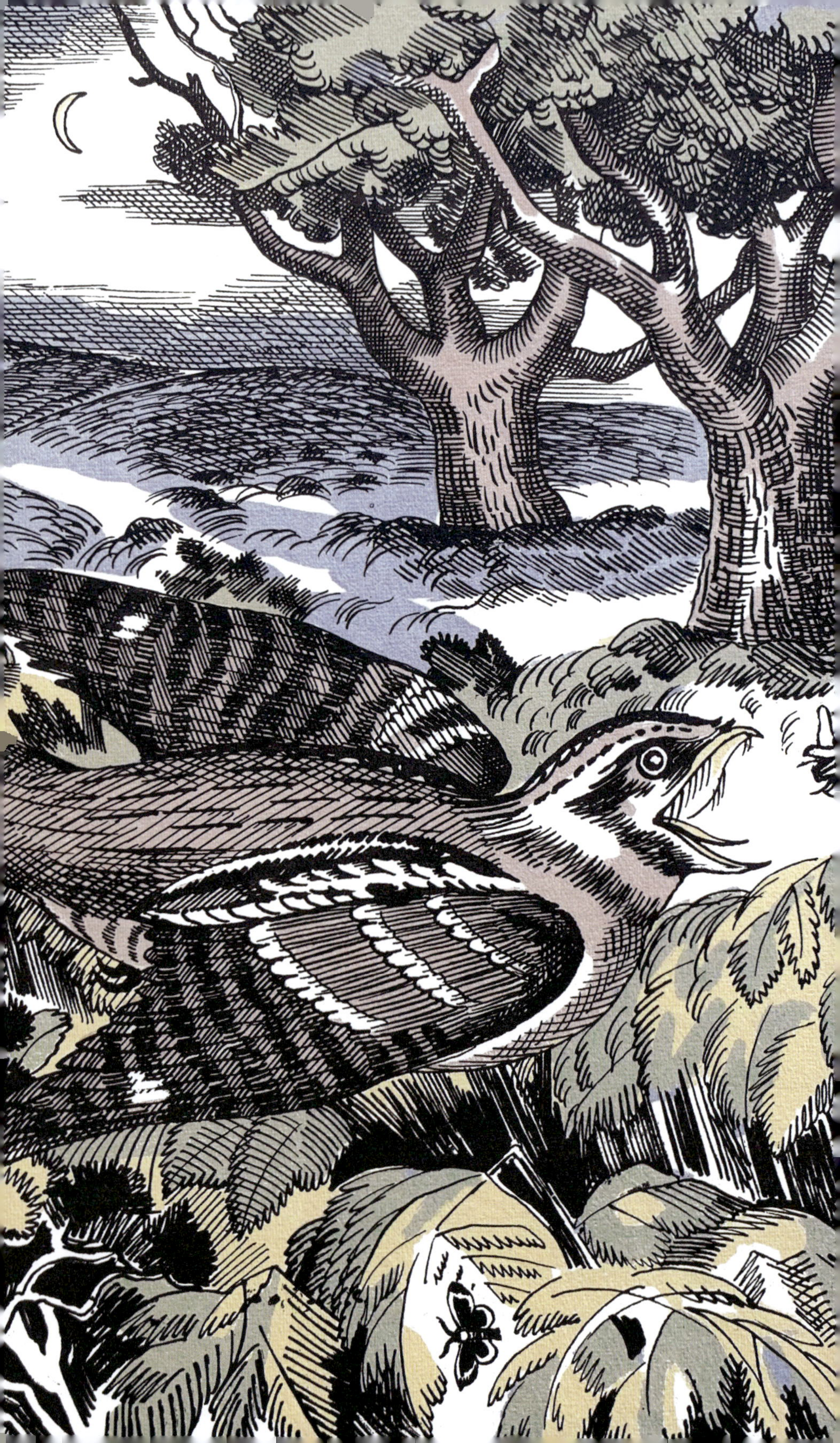

4.
THE MODERN ILLUSTRATORS

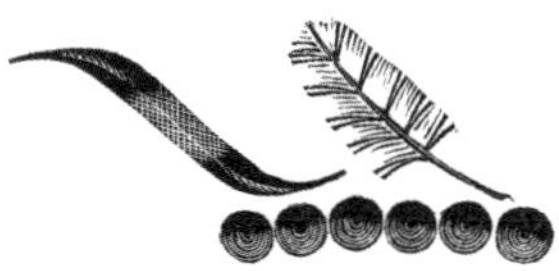

FIG.73
Detail from '*The powers of its wing were wonderful*', 1951/72 (see fig.58).

EDMUND HORT NEW (1871–1931)

1900: Line drawings

The Natural History of Selborne by Gilbert White edited by Grant Allen, London: The Bodley Head, 1900.

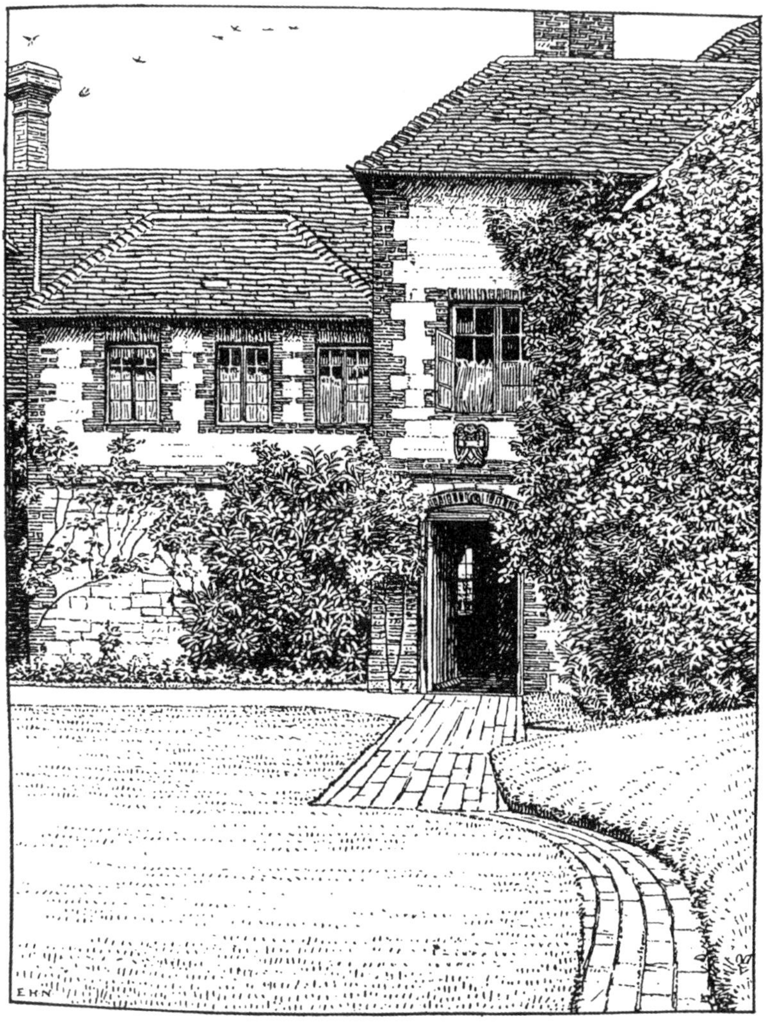

FIG.74

A Corner of White's House.

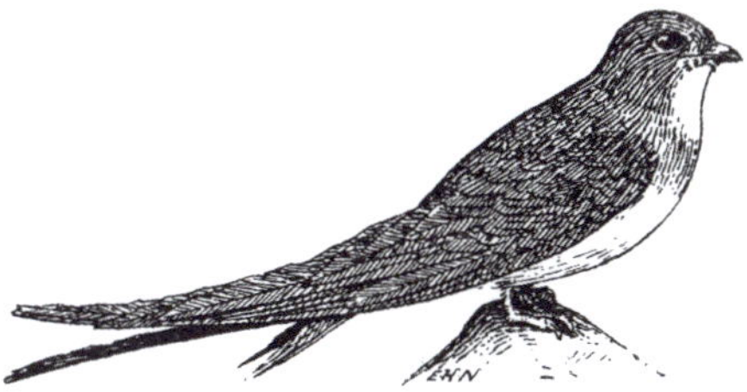

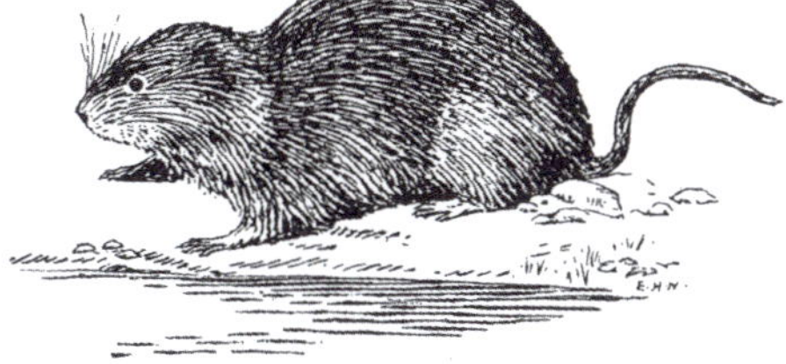

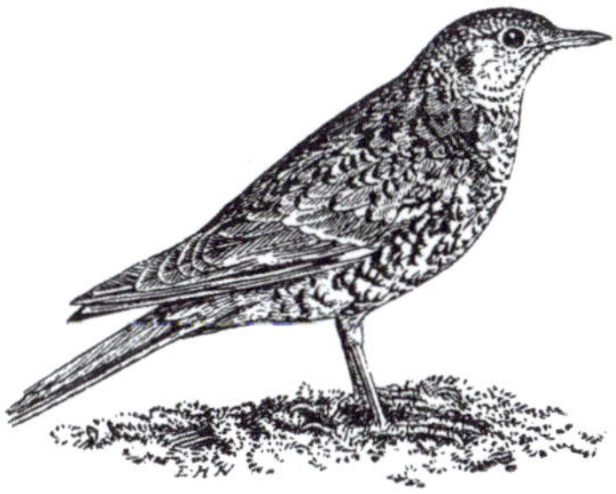

FIG.75
(clockwise from top left):
Gibralter Swift, Water Vole, White's Thrush.

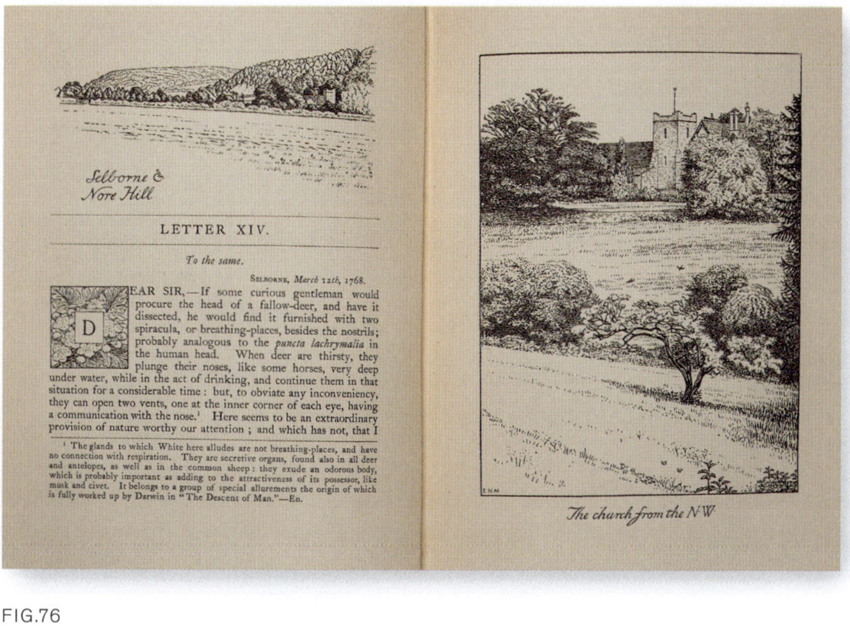

Selborne & Nore Hill

LETTER XIV.

To the same.

SELBORNE, *March 12th*, 1768.

DEAR SIR,—If some curious gentleman would procure the head of a fallow-deer, and have it dissected, he would find it furnished with two spiracula, or breathing-places, besides the nostrils; probably analogous to the *puncta lachrymalia* in the human head. When deer are thirsty, they plunge their noses, like some horses, very deep under water, while in the act of drinking, and continue them in that situation for a considerable time: but, to obviate any inconveniency, they can open two vents, one at the inner corner of each eye, having a communication with the nose.[1] Here seems to be an extraordinary provision of nature worthy our attention; and which has not, that I

[1] The glands to which White here alludes are not breathing-places, and have no connection with respiration. They are secretive organs, found also in all deer and antelopes, as well as in the common sheep: they exude an odorous body, which is probably important as adding to the attractiveness of its possessor, like musk and civet. It belongs to a group of special allurements the origin of which is fully worked up by Darwin in "The Descent of Man."—ED.

The church from the N W

FIG.76
Illustrated spread with the beginning of Letter XIV.

ERIC FITCH DAGLISH (1882–1966)

1929: Wood-engravings

Gilbert White, *The Natural History of Selborne*, E.M. Nicholson, ed., London: Thornton Butterworth Ltd., 1929.

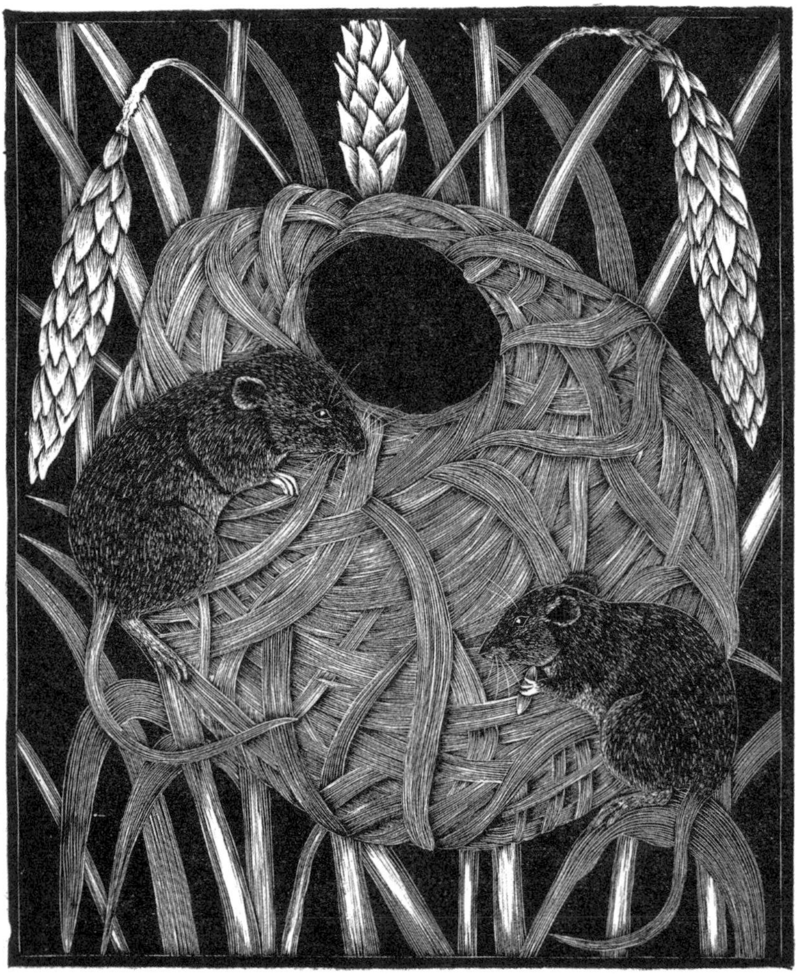

FIG.77
Harvest Mice.

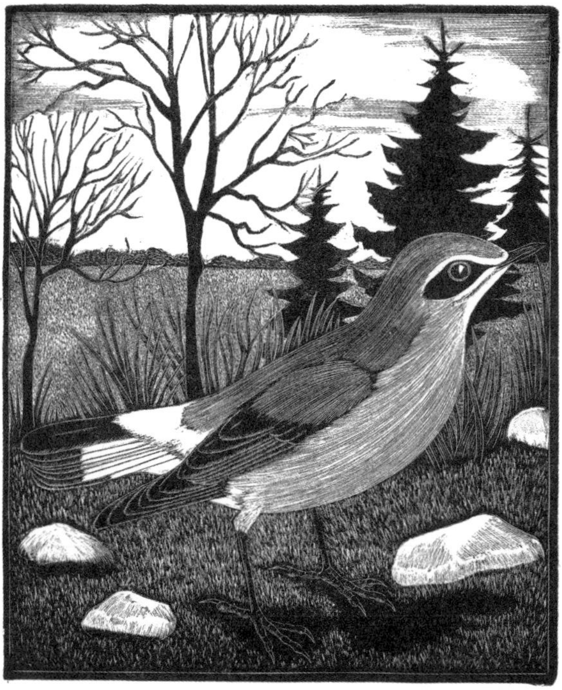

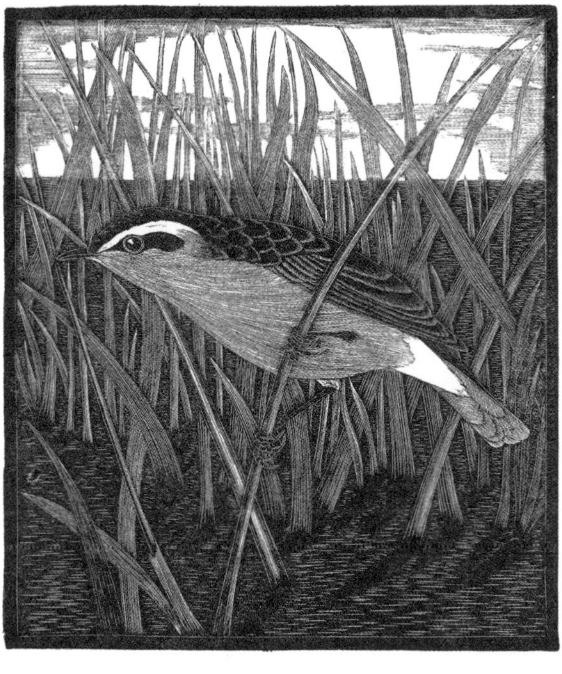

FIG.78
Wheatear.

FIG.79
Sedge Warbler.

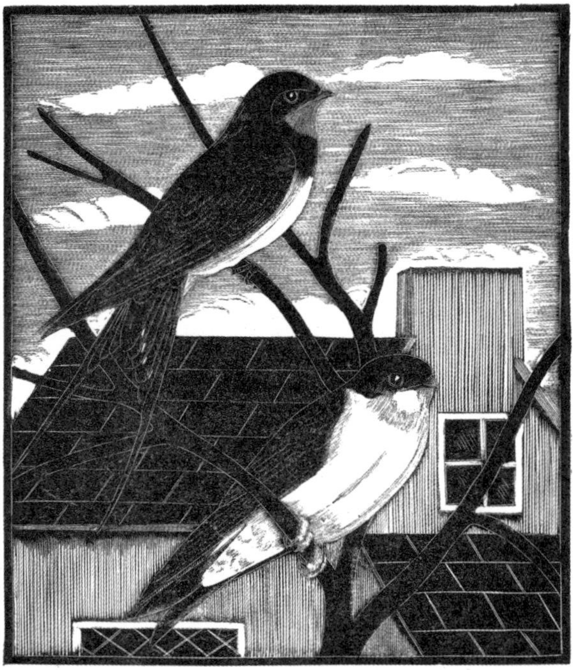

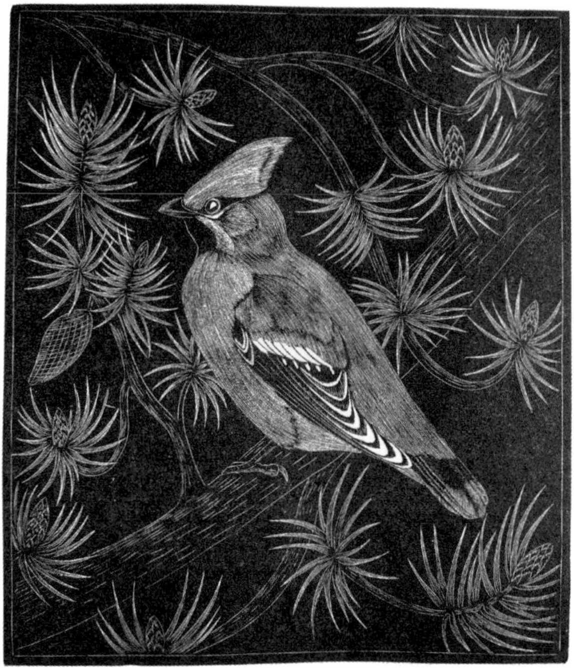

FIG.80
Swallow and Martin.

FIG.81
The German Silk-tail (Waxwing).

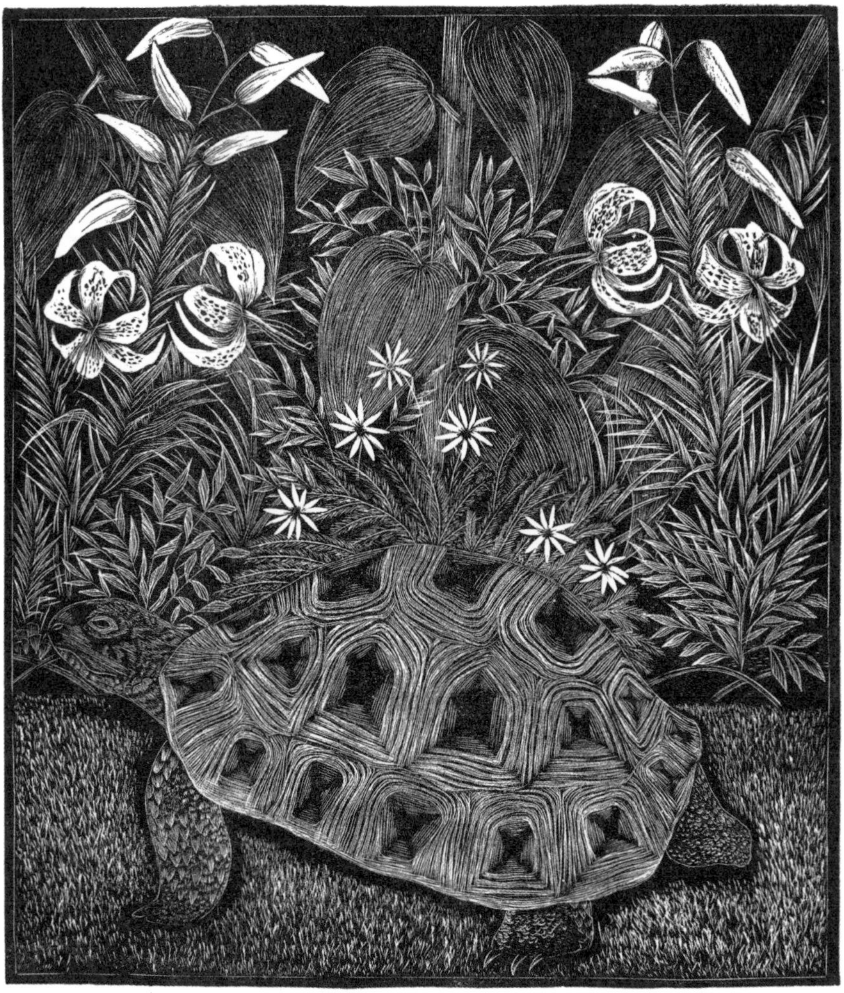

FIG.82
The Tortoise in the Garden.

GERTRUDE HERMES RA (1901–93)

1932: Wood-engravings

William Condry, *Wood Engravings by Gertrude Hermes being Illustrations to Selborne with Extracts from Gilbert White,* Newtown: Gwasg Gregynog, 1988.

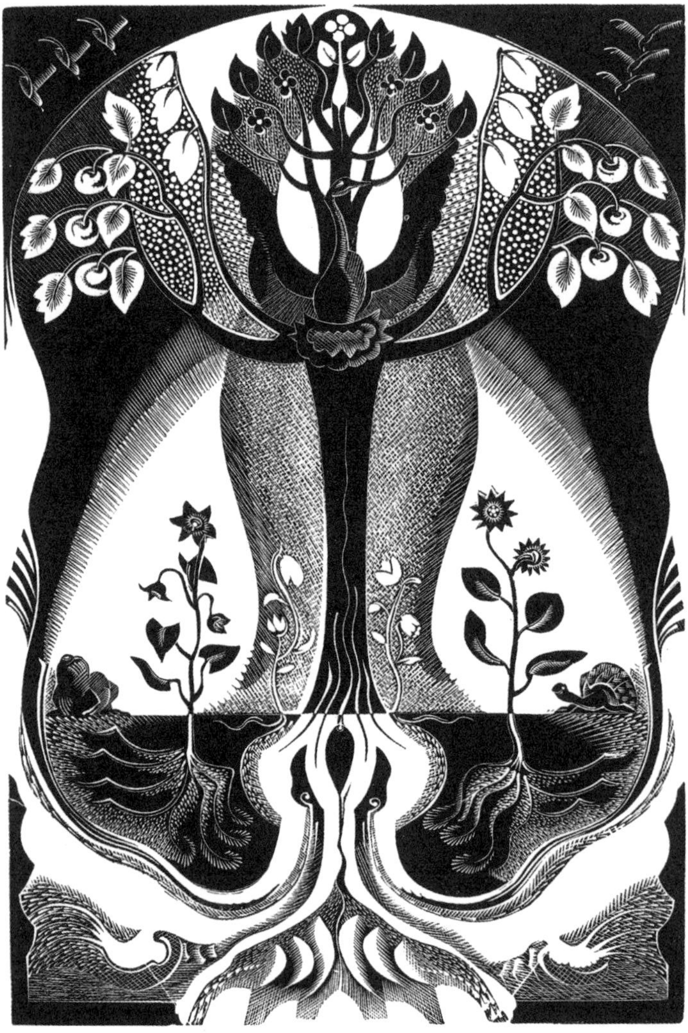

FIG.83
Tree.

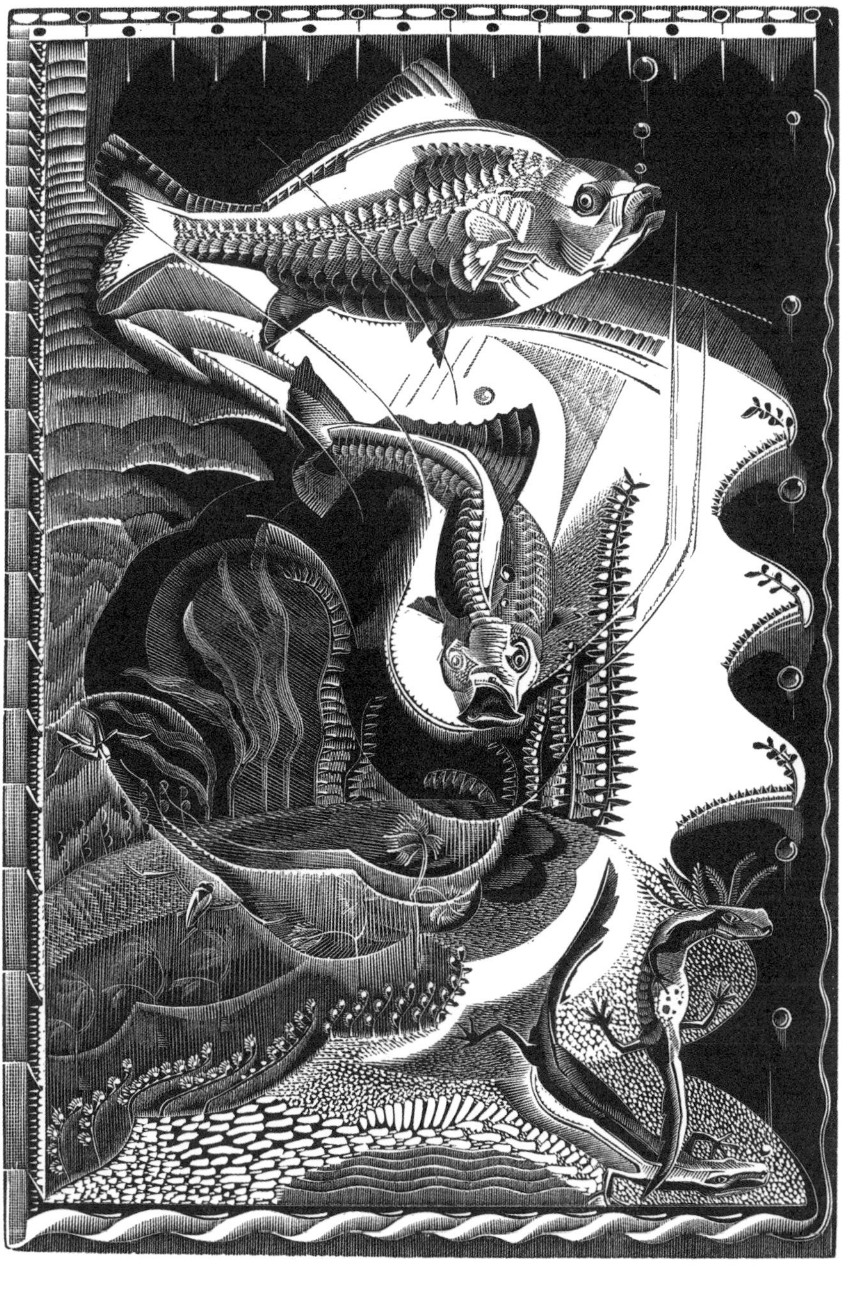

FIG.84
Carp, 1932.

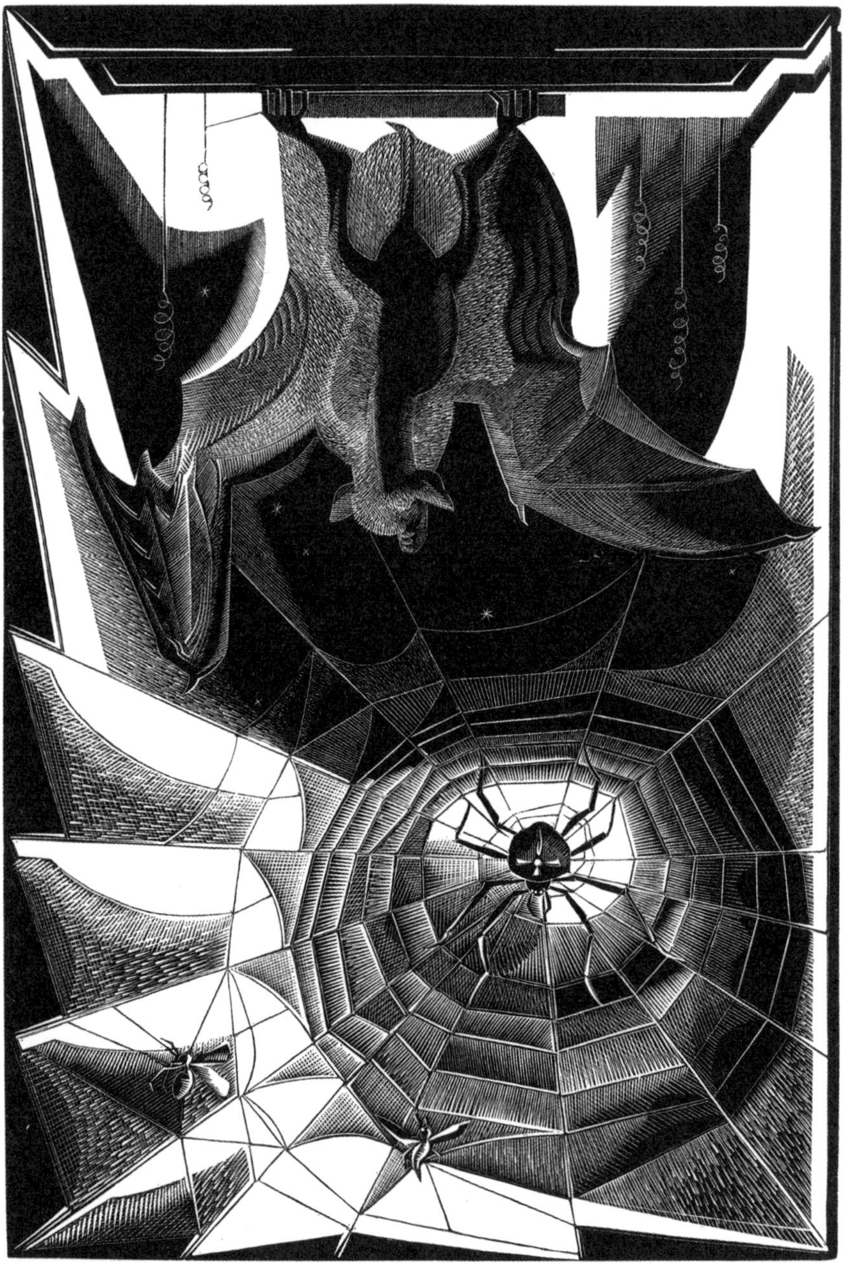

FIG.85
Spider and Bat.

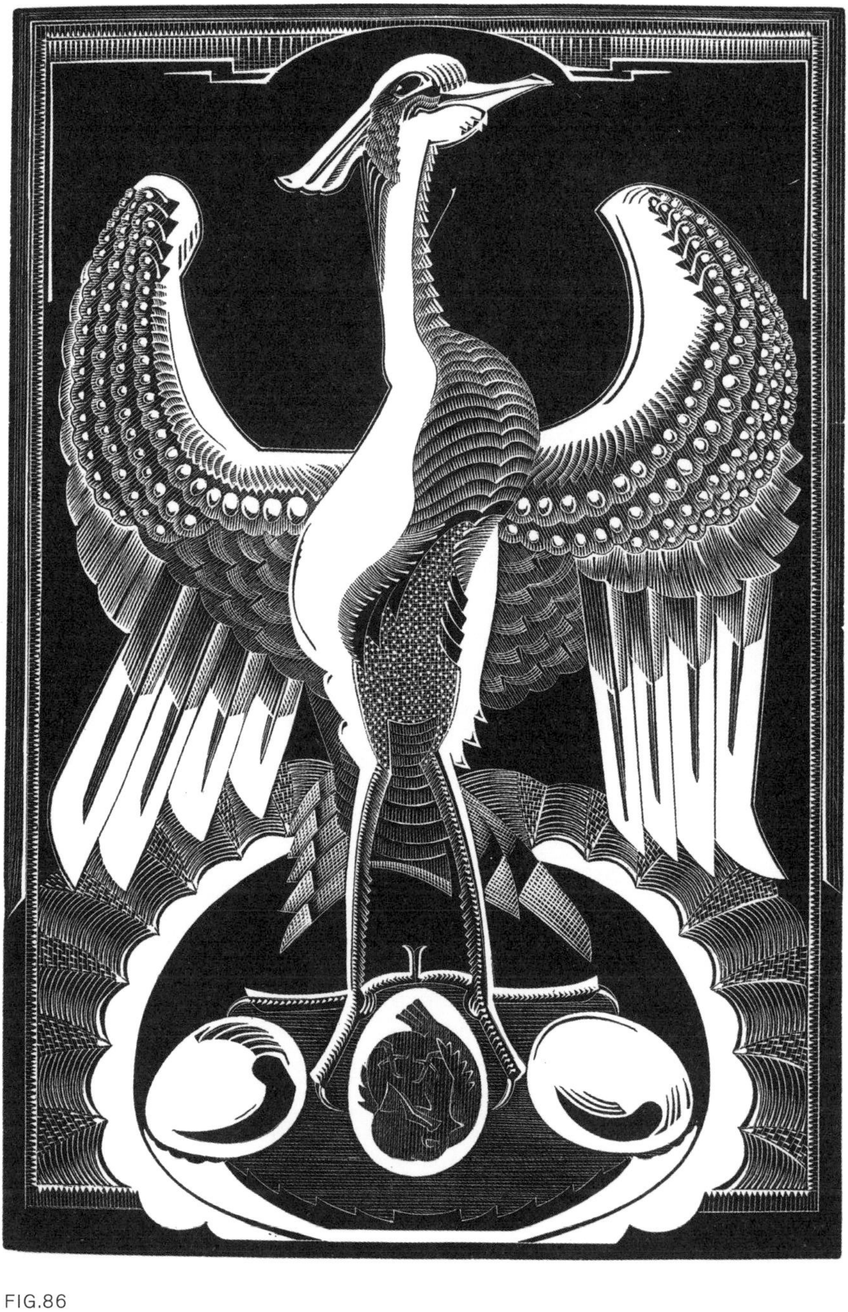

FIG.86
Heron.

ERIC RAVILIOUS (1903–42)

1938: Wood-engravings

The Writings of Gilbert White, H.J. Massingham, ed., London: The Nonesuch Press, 1938.

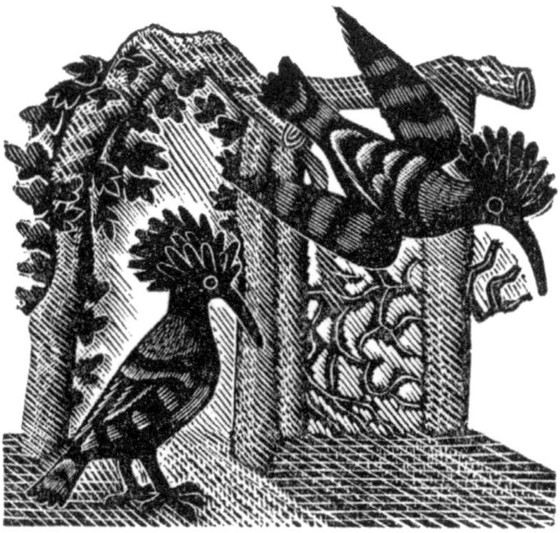

FIG.87
Hoopoes, tailpiece to volume two.

FIG.88
The Tortoise and *The Pheasant.*

FIG.89 (opposite)
Title page for volume two.

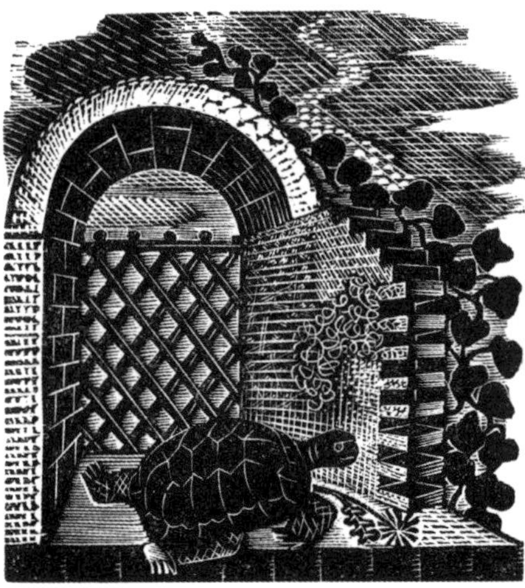

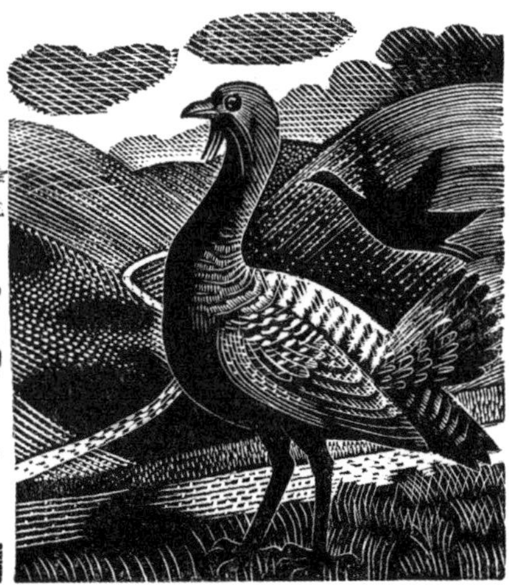

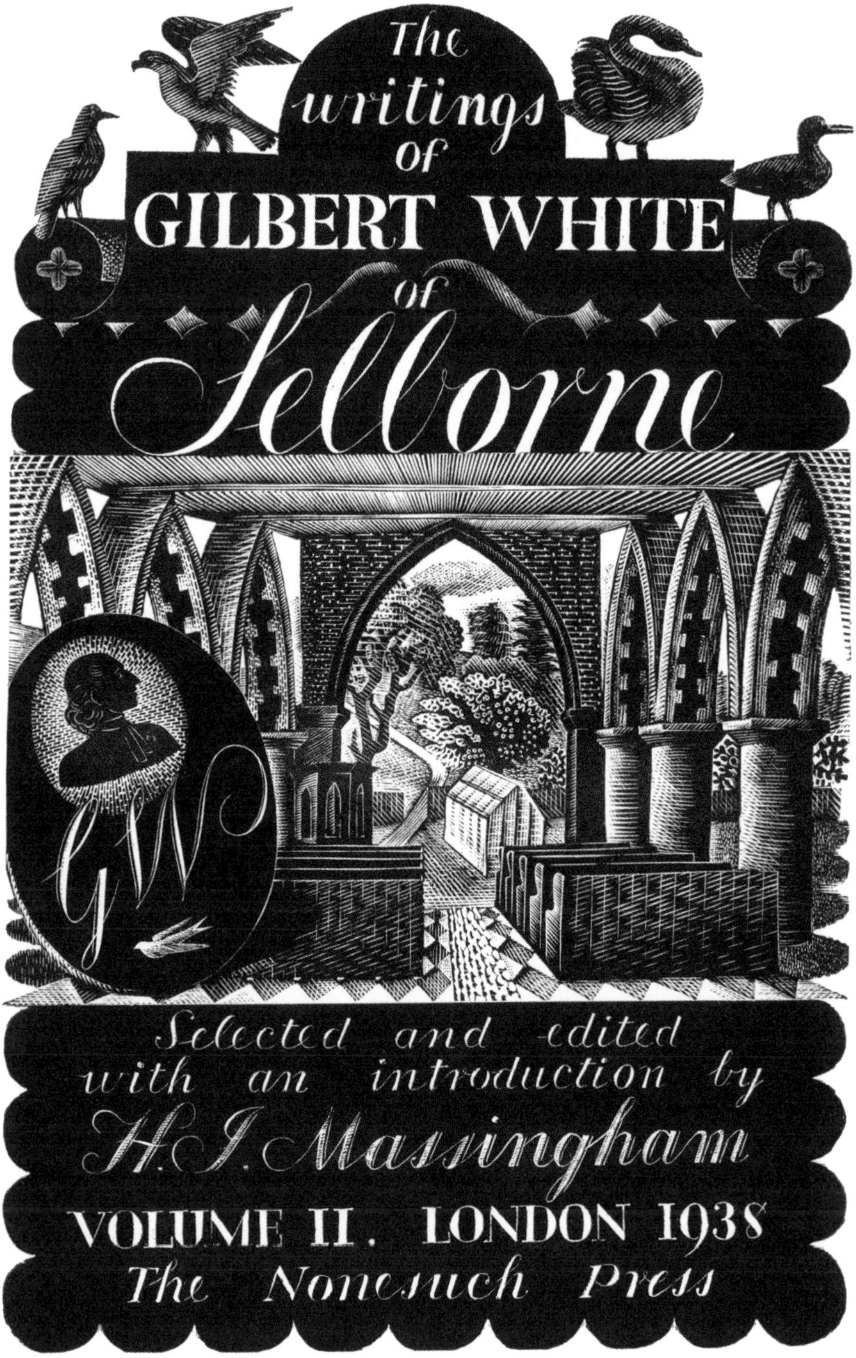
The writings of
GILBERT WHITE
of
Selborne
Selected and edited with an introduction by
H. J. Massingham
VOLUME II. LONDON 1938
The Nonesuch Press

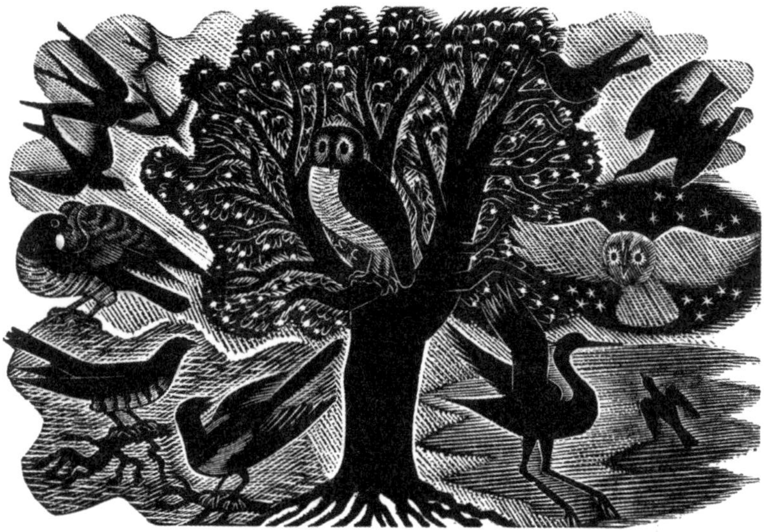

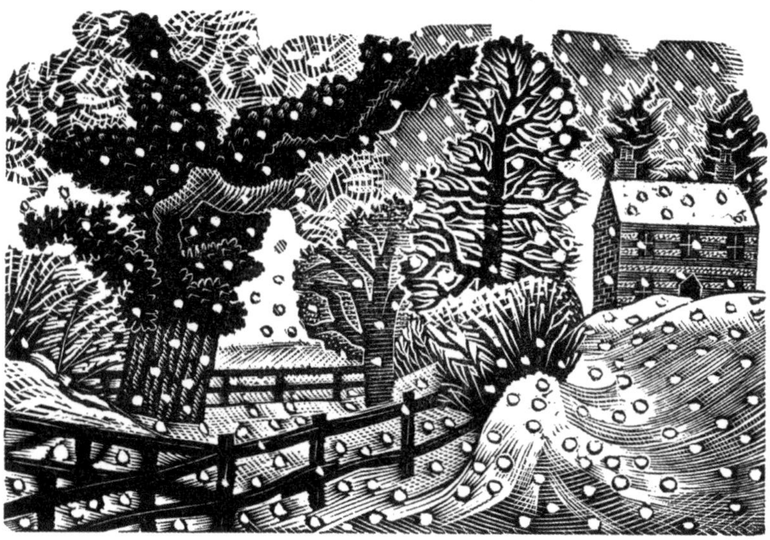

FIG.90 (above)
Requirements of an Ornithologist.

'A good ornithologist should be able to distinguish birds by their air as well as by their colours and shape; on the ground as well as on the wing, and in the bush as well as in the hand.' Letter XLII, 7 August 1778

FIG.91 (below)
'Considerable falls of snow'.

'For the last two or three days of the former year there were considerable falls of snow, which lay deep and uniform on the ground without any drifting, wrapping up the more humble vegetation in perfect security.' Letter LXI to the Hon. Daines Barrington.

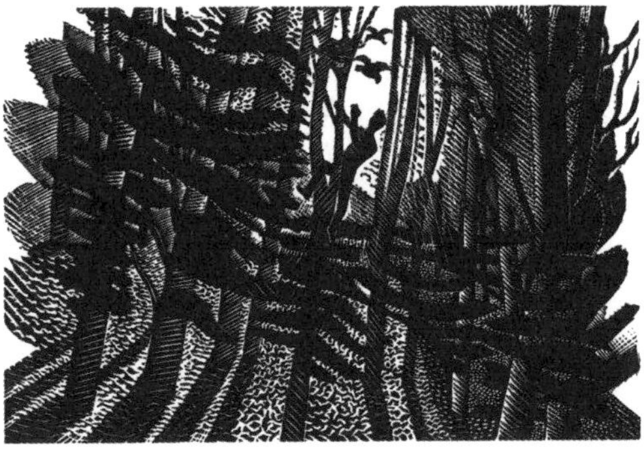

FIG.92
Birdsnesting.

'A pair of honey-buzzards, built them a large shallow nest,... upon a tall slender beech near the middle of Selborne hanger... A bold boy climbed this tree, though standing on so steep and dizzy a situation, and brought down an egg.'

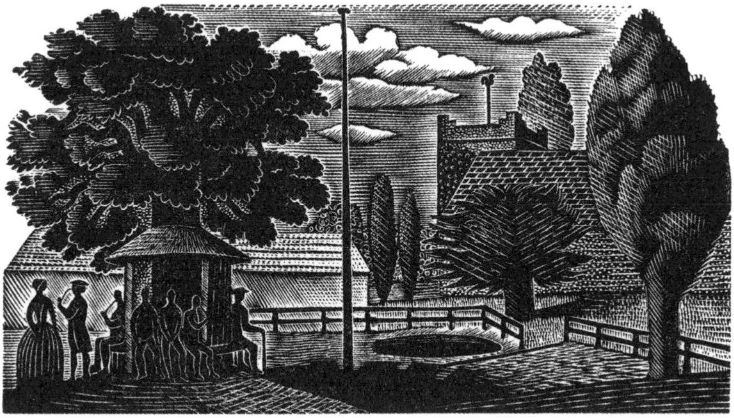

FIG.93
The Plestor.

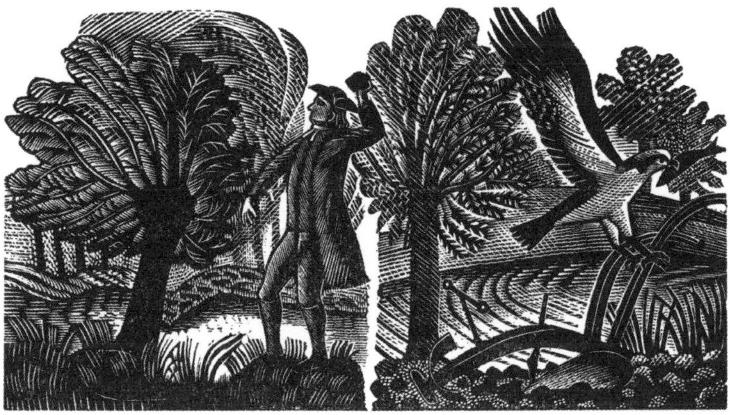

FIG.94
The Sedge-bird and *The Osprey.*

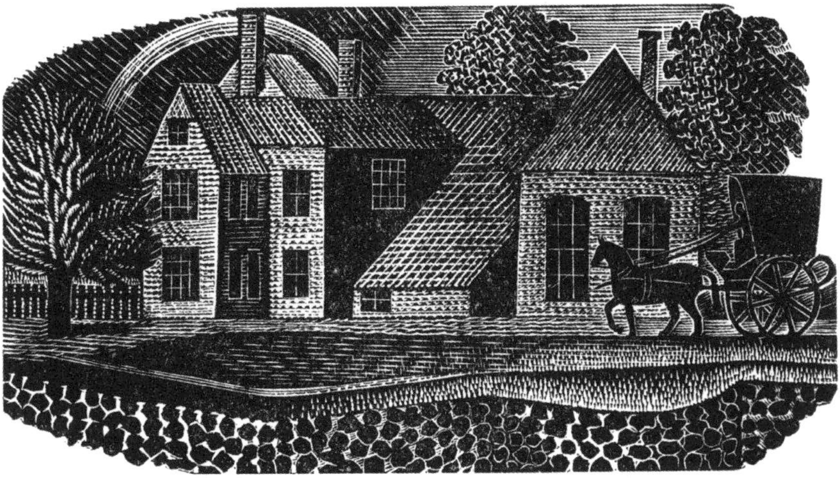

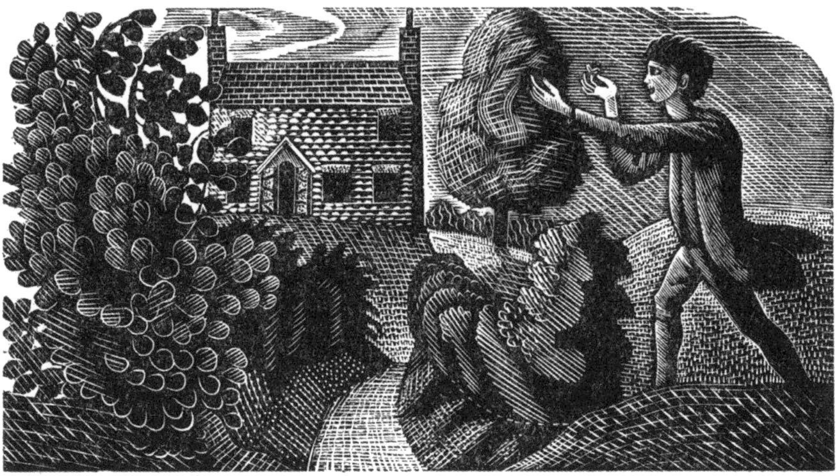

FIG.95 (above)
East Sussex House.

This is presumably intended to represent Delves House on Ringmer Green, the home of White's aunt Mrs Rebecca Snooke, from whom he acquired his tortoise.

FIG.96 (below)
The Idiot Boy.

'We had in this village more than twenty years ago an idiot-boy, whom I well remember, who, from a child, shewed a strong propensity to bees; they were his food, his amusement, his sole object.'

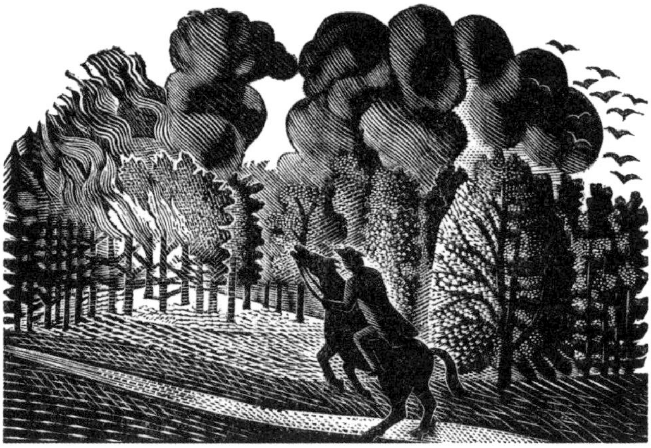

FIG.97
Woolmer Forest.

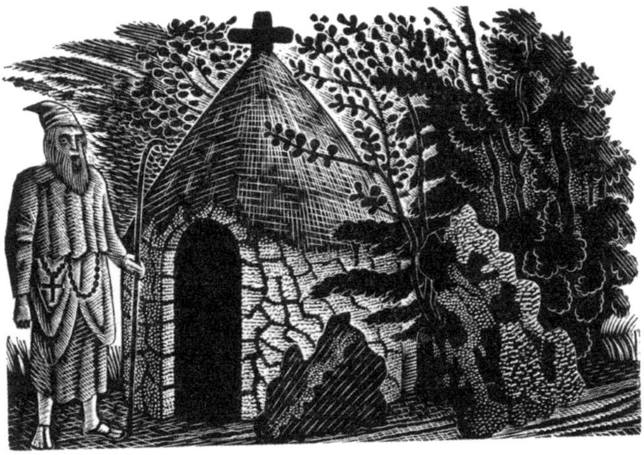

FIG.98
The Hermit.

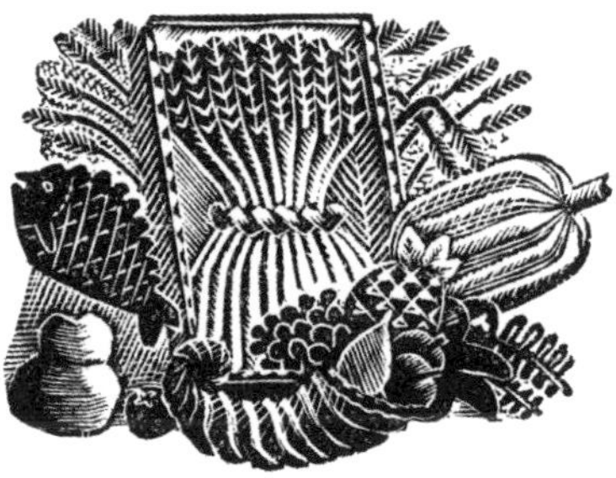

FIG.99
Tailpiece.

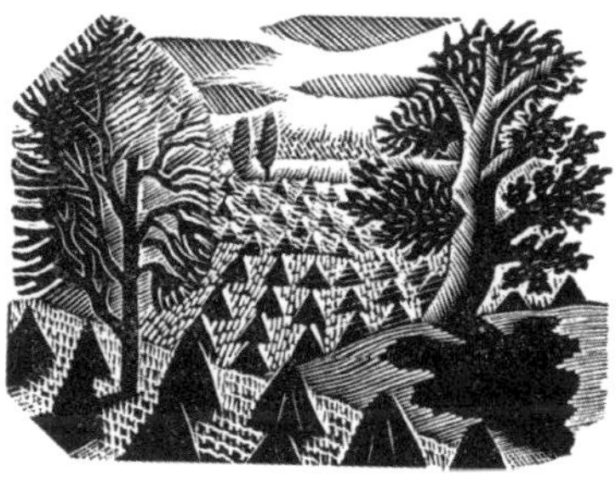

FIG.100
Tailpiece.

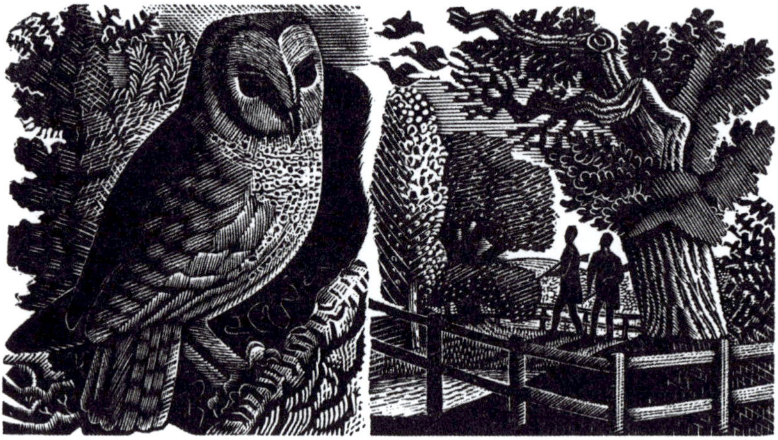

FIG.101
The White Owl and *The Hunters.*

'Some young men went down lately to a pond on the verge of Wolmer-forest to hunt flappers, or young wild ducks, many of which they caught, and among the rest, some very minute yet well-fledged wild-fowls alive, which upon examination I found to be teals. I did not know until then that teals ever bred in the south of England, and was very much pleased with the discovery: this I look upon as a great stroke in natural history. We have had, ever since I can remember, a pair of white owls that constantly breed under the eaves of this church.' Letter XV to the Hon. Daines Barrington, 8 July 1773.

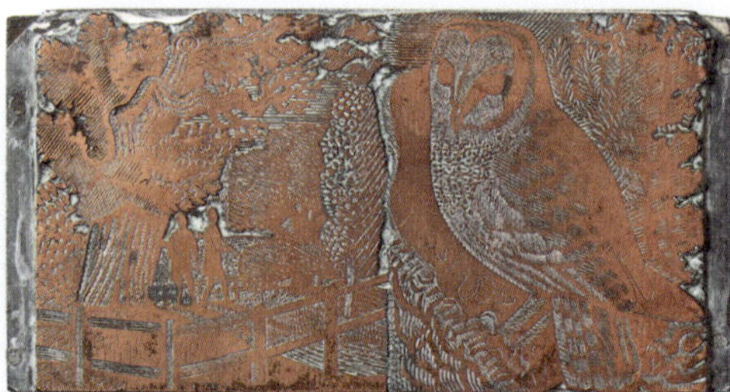

FIG.102
Electrotype block for *The White Owl* and *The Hunters*, 1938 (see fig.101).

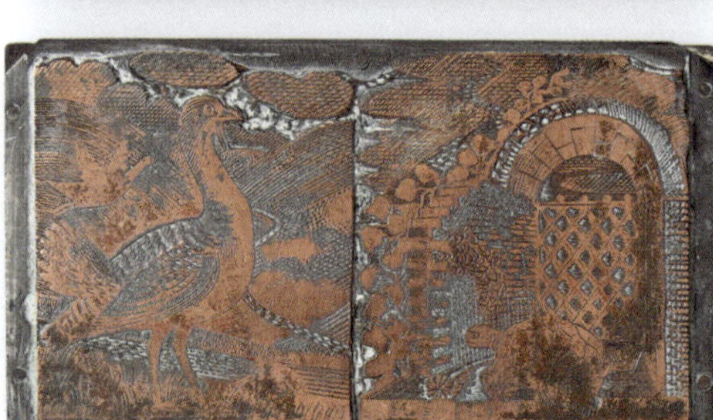

FIG.103
Electrotype block for *The Tortoise* and *The Pheasant*, 1938 (see fig.88).

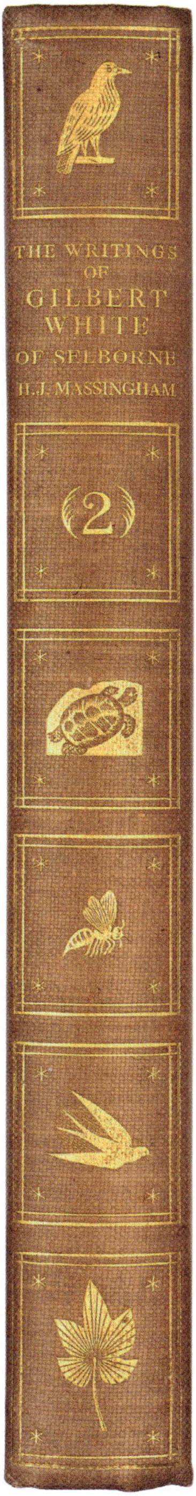

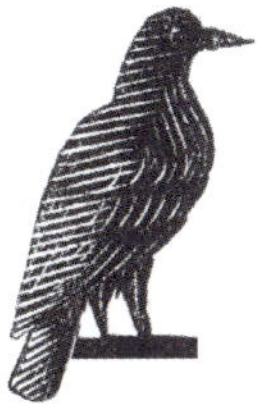

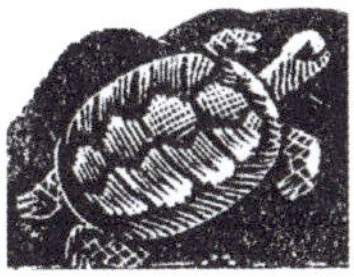

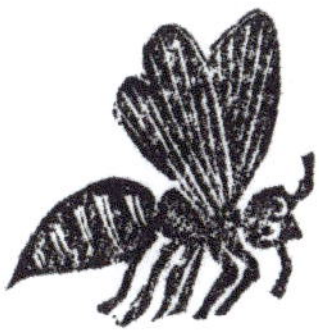

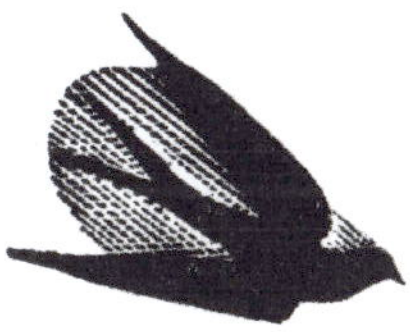

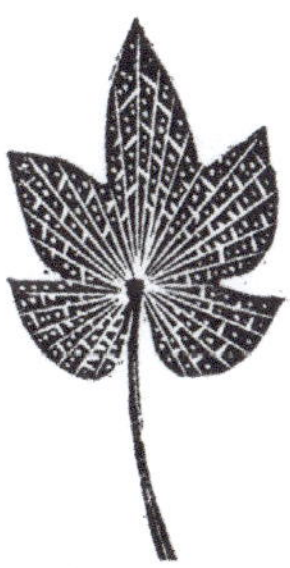

FIG.104 (left)
Spine of volume two, showing five decorative panels from Ravilious's wood-engravings.

FIG.105
Raven, top left of title page, volumes one and two (see figs. 9 and 89).

FIG.106
Tortoise, bottom left of *The Tortoise in the Kitchen Garden* (see fig.39).

FIG.107
Wasp, bottom left of tailpiece of a pear (striped vert longue) (see p.189).

FIG.108
Swallow, top left of *Requirements of an Ornithologist* (see fig.90).

FIG.109
Leaf, top right of tailpiece, p.viii, volume two.

AGNES MILLER PARKER (1895–1980)

1941: wood-engravings

Leonard Russell, ed., *The Saturday Book 1941–2: A New Miscellany*, London: Hutchinson, 1941.

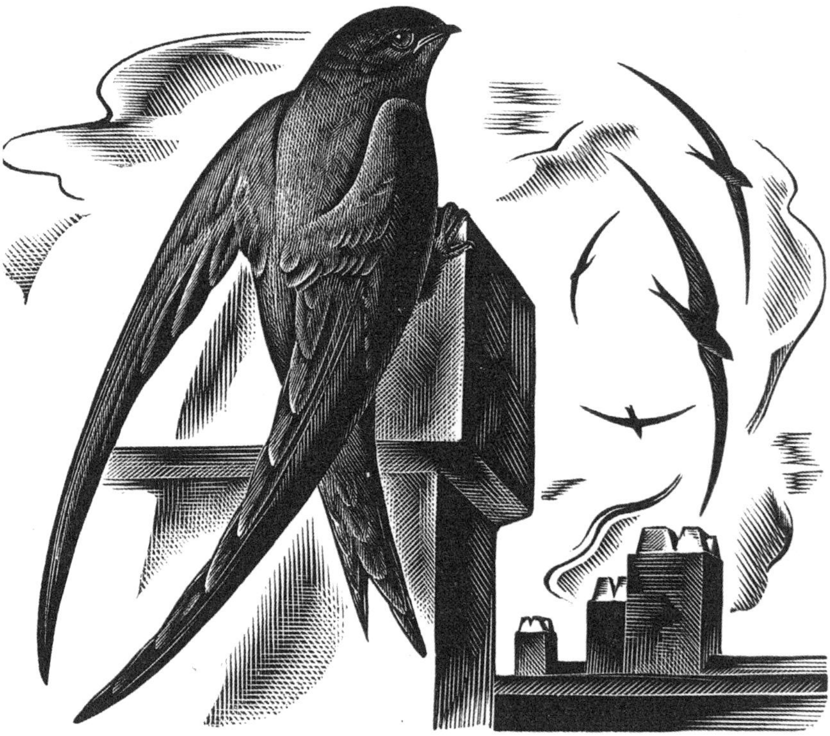

FIG.110
Swift.

'Swifts, as I suspected, invariably lay but *two* eggs; and as they breed but *once*, their encrease is very small! I got Harry's bricklayer one evening to open the tiles of his bre-house, under which were several nests containing only *two* squab young apiece; and moreover his workmen all told me that, when boys they had invariably found only two eggs or two birds. If I lived at Fyfield I should be more learned in swifts; for as you sit in the parlour, you see their proceedings at the brewhouse.'

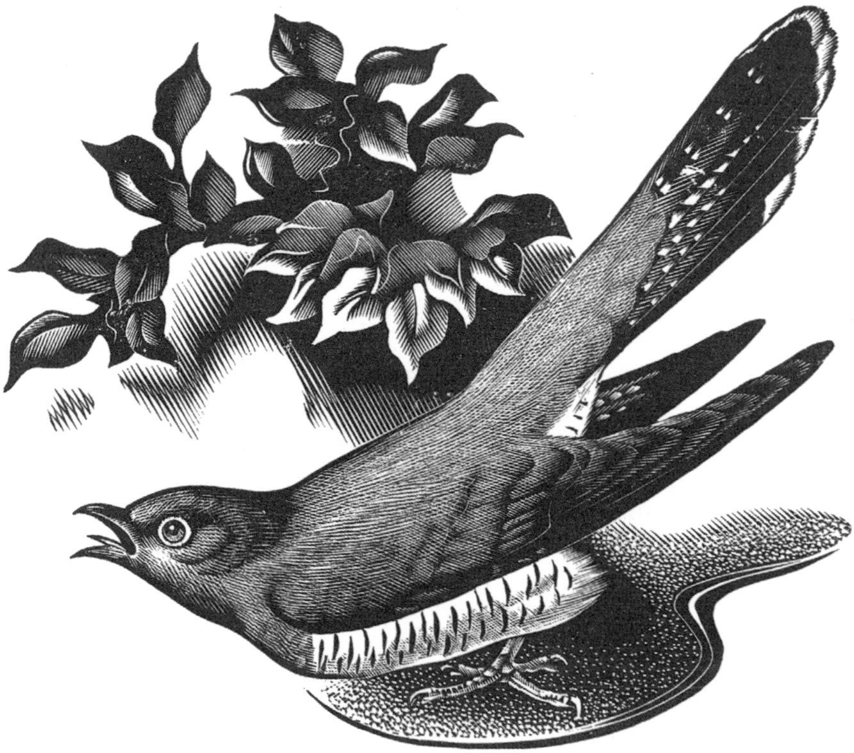

FIG. 111
Cuckoo.

'In July, I saw several cuckoos skimming over a large pond; and found, after observation, that they were feeding on the libellulæ, or *dragon-flies*; some of which they caught as they settled on the weeds, and some as they were on the wing. Notwithstanding what Linnæus says, I cannot be induced to believe that they are birds of prey.'

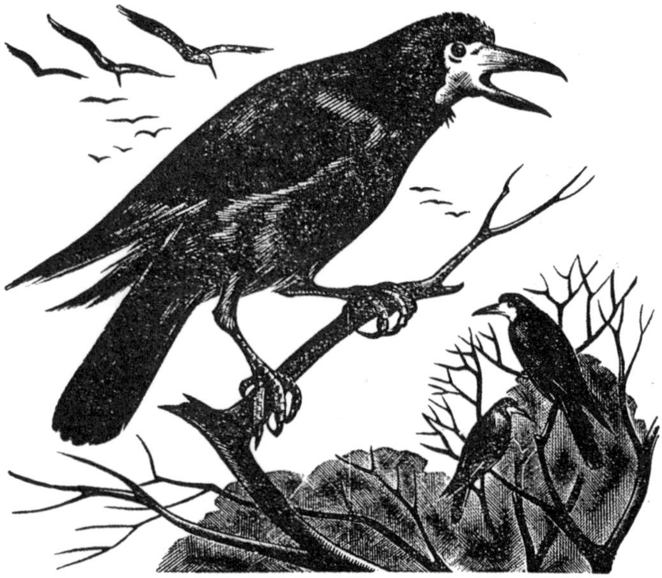

FIG. 112
Rooks.

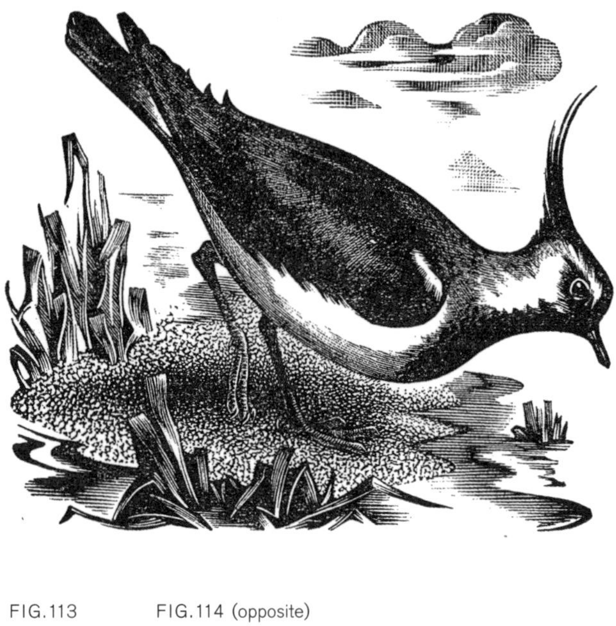

FIG. 113
Lapwing.

FIG. 114 (opposite)
Starlings.

BIRDS

Agnes Miller Parker

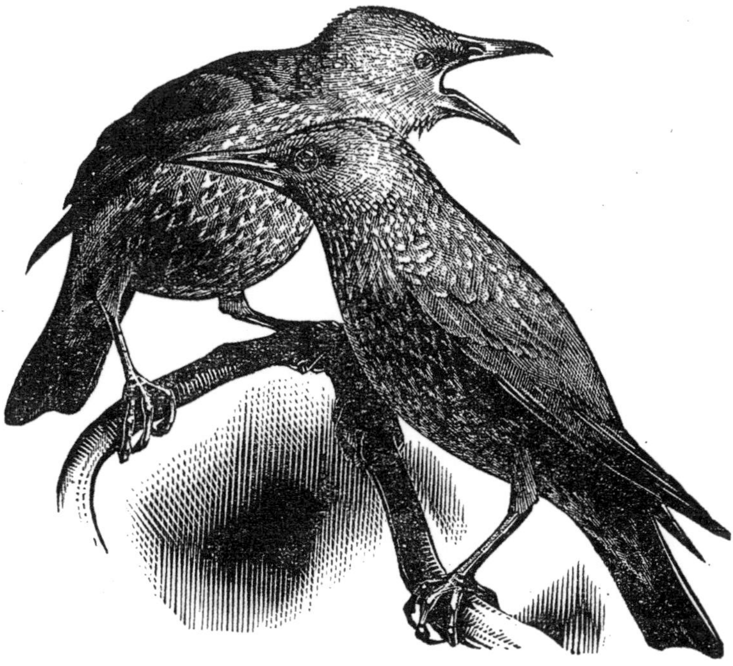

STARLINGS

Starlings as it were swim along, while *missel-thrushes* use a wild and desultory flight. . . .

GILBERT WHITE[1]

1 This extract and the others which follow are from the Nonesuch edition of Gilbert White, 1937, edited by H. J. Massingham.

CLARE LEIGHTON

1941: wood-engravings

Gilbert White, *The Natural History of Selborne*, Harmondsworth: Penguin Books, 1941.

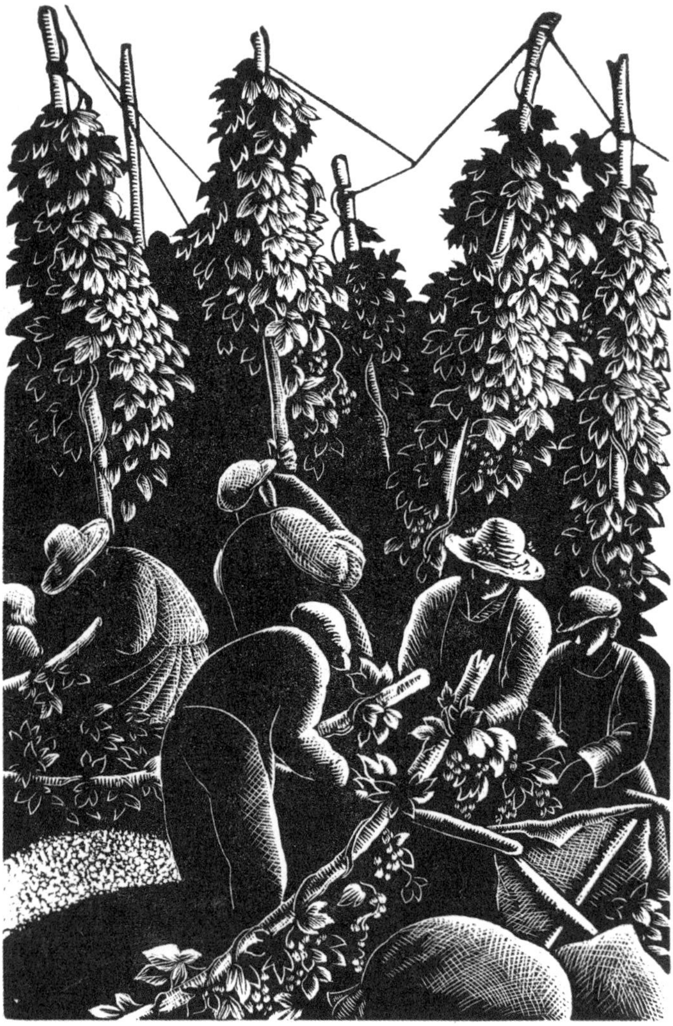

FIG.115
Hop Picking.

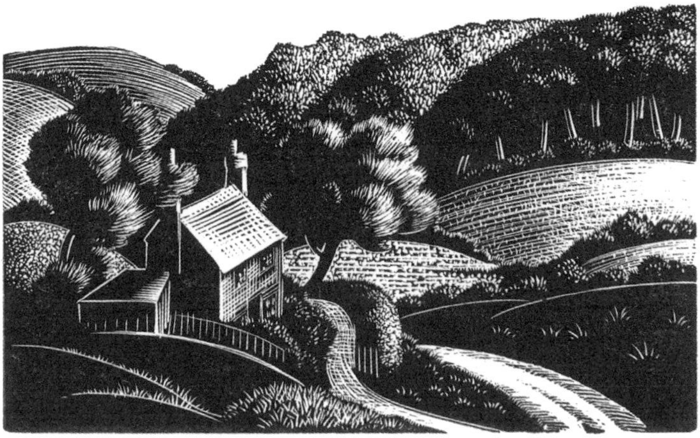

FIG.116
Landscape. Illustration to Letter I to Thomas Pennant.

This image depicts Huckers Cottage near the Short Lithe and Selborne Church (also seen in an earlier form in Samuel Hieronymus Grimm's *North East View of Selborne, from the Short Lythe*; fig.14).

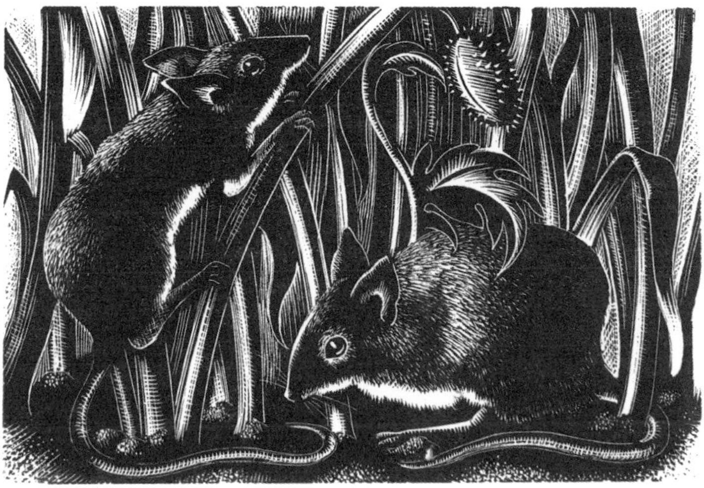

FIG.117
Field Mice.

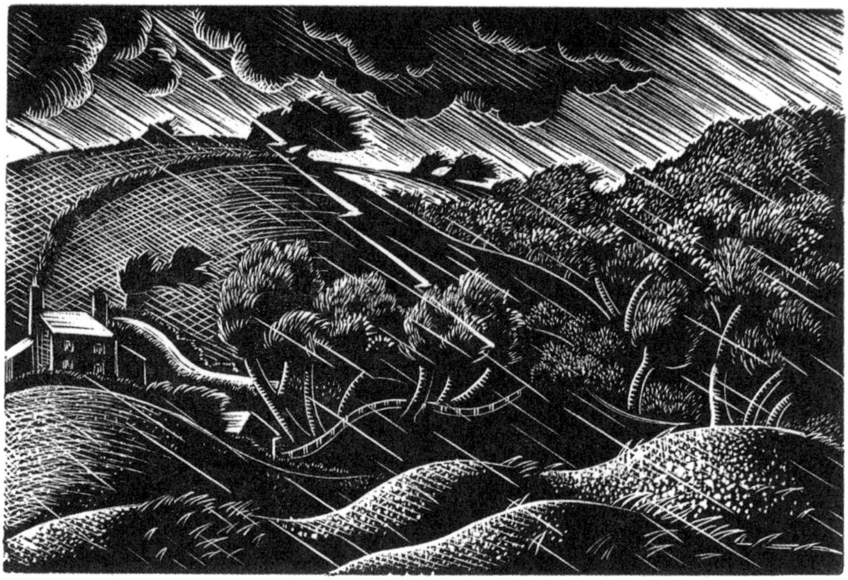

FIG. 118
The Storm. Illustration to Letter XXIX to the Hon. Daines Barrington, Selborne, 12 May 1770.

'Last month we had such a series of cold, turbulent weather, such a constant succession of frost, and snow, and hail, and tempest that the regular migration or appearance of the summer birds was much interrupted. Some did not show themselves (at least were not heard) till weeks after their usual time; as the *blackcap* and *whitethroat*; and some have not been heard yet, as the *grasshopper-lark* and largest *willow-wren*. As to the *fly-catcher*, I have not seen it: it is indeed one of the latest, but should appear about this time: and yet, amidst all this meteorous strife and war of the elements, two swallow discovered themselves as long ago as *April* 11th in frost and snow; but they withdrew quickly, and were not visible again for many days.'

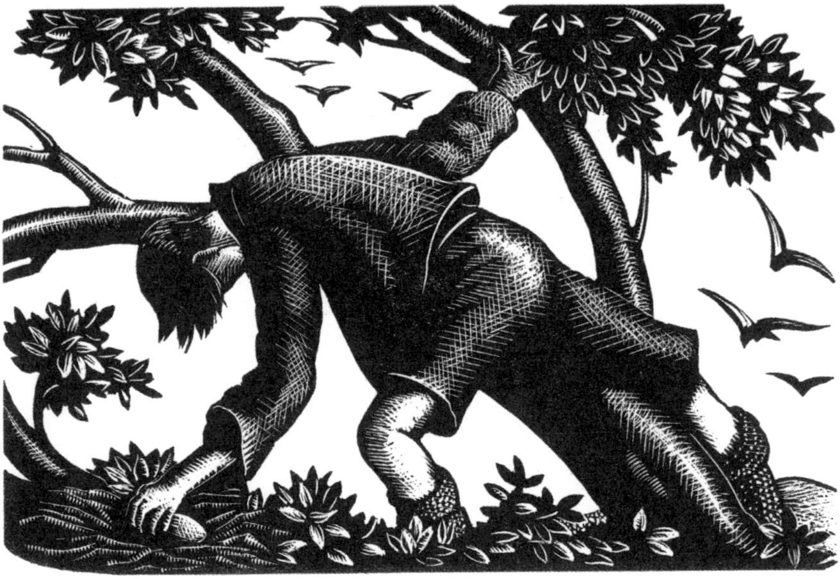

FIG.119
Birdsnesting. Illustration to Letter XLIII to Thomas Pennant.

'A pair of honey-buzzards, *Buteo apivorus* Linn., sive *Vespivorus* Raii, built them a large shallow nest, composed of twigs and lined with dead beechen leaves, upon a tall slender beech near the middle of Selborne Hanger in the summer of 1780. In the middle of the month of June a bold boy climbed this tree, though standing on so steep and dizzy a situation, and brought down an egg, the only one in the nest, which had been sat on for some time, and contained the embryo of a young bird. The egg was smaller, and not so round as those of the common buzzard; was dotted at each end with small red spots, and surrounded in the middle with a broad bloody zone.'

FIG.120
Osprey. Illustration to Letter XXXIX to Thomas Pennant, 9 November 1773.

'The osprey was shot a year ago at Frinsham Pond, a great lake, at about six miles from hence, while it was sitting on the handle of a plough and devouring a fish: it used to precipitate itself into the water, and so take its prey by surprise.'

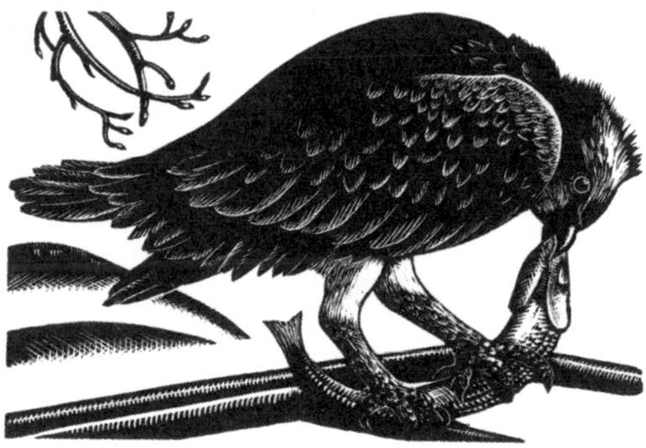

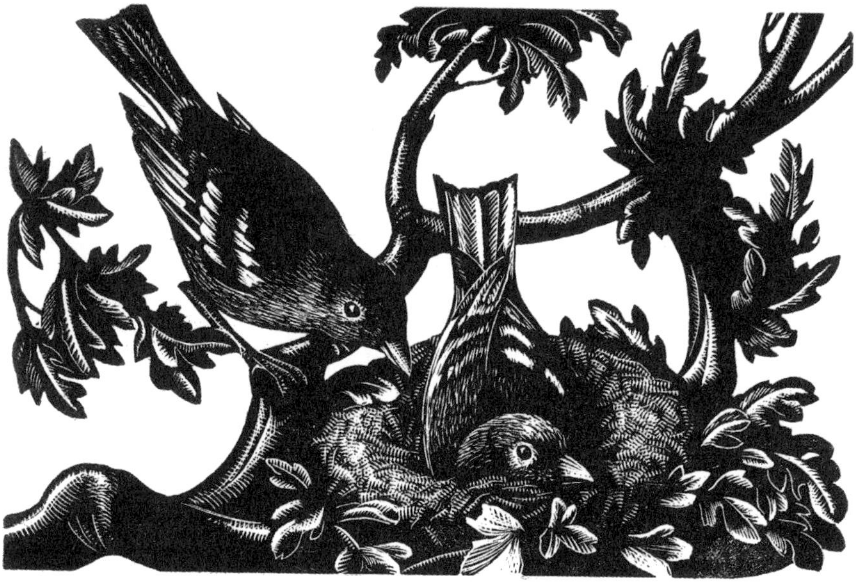

FIG.121
Chaffinches.

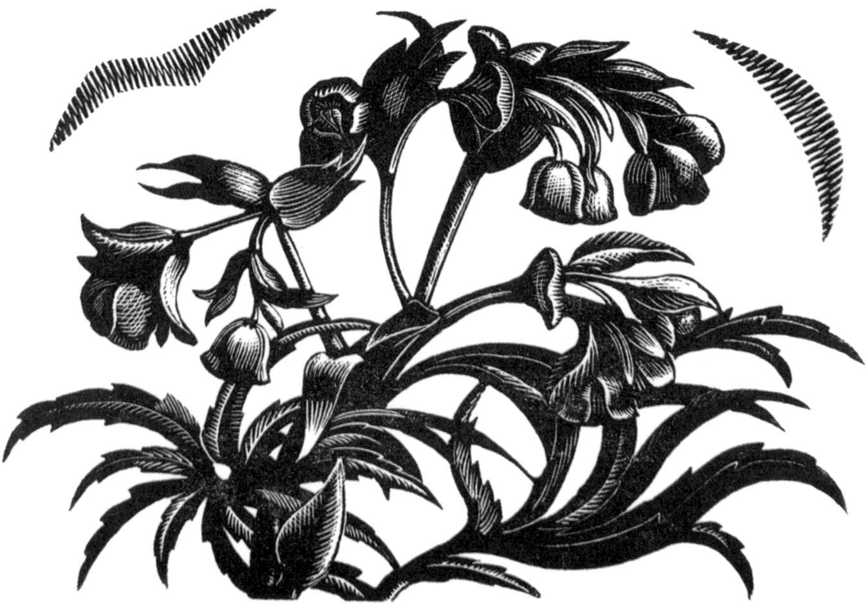

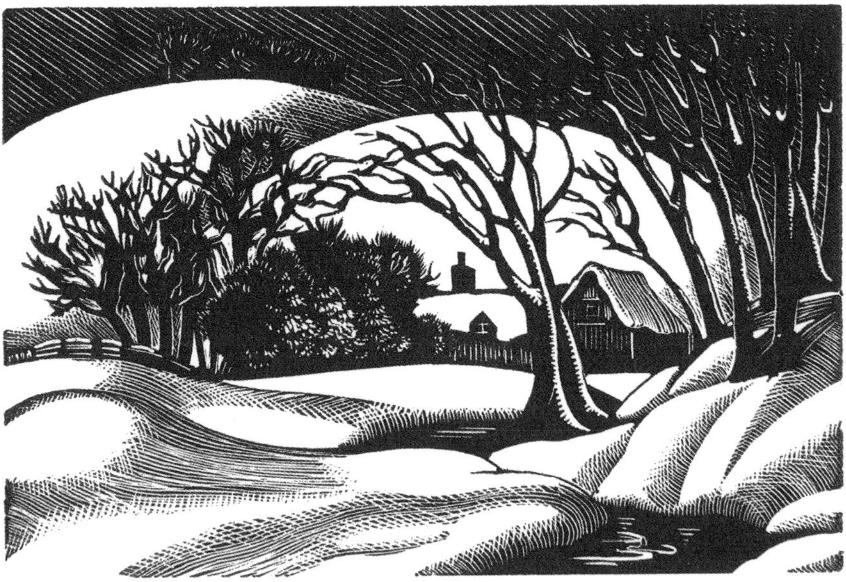

FIG.122 (opposite, below)
Stinking Hellebore. Illustration to Letter XLI to the Hon. Daines Barrington, 3 July 1778.

'*Helleborus fœtidus*, stinking hellebore, bear's foot, or setterworth, - all over the High-wood and Coney-croft-hanger: this continues a great branching plant the winter through, blossoming around January, and is very ornamental in shady walks and shrubberies. The good women give the leaves powered to children troubled with worms; but it is a violent remedy, and ought to be administered with caution.'

FIG.123
Snow Scene. Illustration to Letter LXII to the Hon. Daines Barrington.

'On the 14th the writer was obliged to be much abroad; and thinks he never before or since has encountered such rugged Siberian weather. Many of her narrow roads were now filled above the tops of hedges, through which the snow was driven into most romantic and grotesque shapes, so striking to the imagination as not to be seen without wonder and pleasure.'

CLAIRE OLDHAM (1906–96)

1947: Wood-engravings

Gilbert White, *The Natural History of Selborne Illustrated by Claire Oldham*, James Fisher, ed., London: Cresset Press, 1947.

FIG.124
South Downs. Illustration to Letter XVII, Ringmer near Lewes, 9 December 1773.

'Though I have now travelled the Sussex Downs upwards of thirty years, yet I still investigate that chain of majestic mountains with fresh admiration year by year; and I think I see new beauties every time I traverse it. This range, which runs from Chichester eastward as far as Eastbourne, is about sixty miles in length, and is called the South Downs, properly speaking, only round Lewes. As you pass along you command a noble view of the wild, or weald, on one hand, and the broad downs and sea on the other.'

FIG.125
Duck and Fox. Illustration to Letter VIII.

'On the verge of the forest, as it is now circumscribed, are three considerable lakes, two in Oakhanger, of which I have nothing particular to say; and one called Bin's, or Bean's Pond, which is worthy the attention of a naturalist or a sportsman. For, being crowded at the upper end with willows, and with the *carex cespitosa*, it affords such a safe and pleasing shelter to wild ducks, teals, snipes, etc., that they breed there. In the winter this covert is also frequented by foxes, and sometimes by pheasants and the bogs produce many curious plants.'

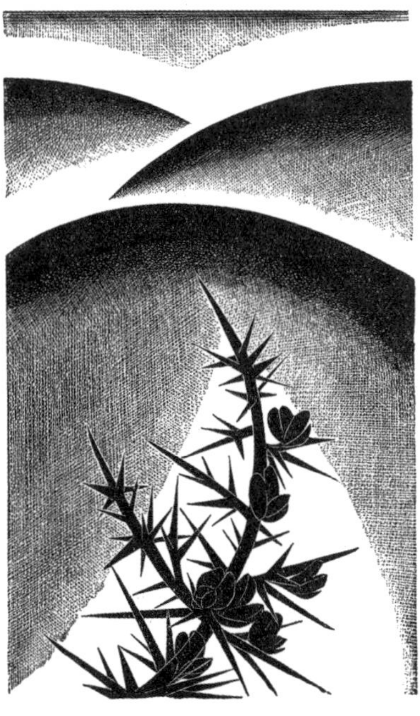

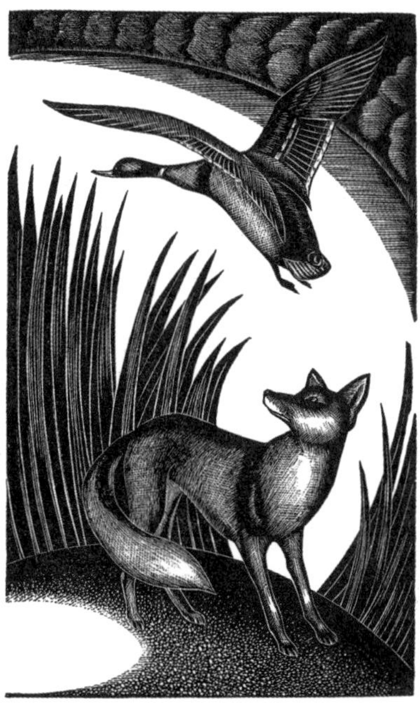

FIG.126
Squirrel, Field Mouse and Nuthatch.
Illustration to Letter LVI.

'There are three creatures, the squirrel, the field-mouse and the bird called the nut-hatch (*Sitta Europœa*), which live much on the hazel-nut; and yet they open them each in a different way. The first, after rasping off the small end, splits the shell in two with his long fore-teeth, as a man does with his knife; the second nibbles a hole with his teeth, so regular as if drilled with a wimble, and yet so small that one could wonder how the kernel can be extracted through it; while the last picks an irregular ragged hole with its bill; but as this artist has no paws to hold the nut firm while he pierces it, like an adroit workman, he fixes it, as it were in a vice, in some cleft of a tree, or in some crevice; when, standing over it, he perforates the stubbon shell.'

FIG.127
Cuckoo and Nightjar. Illustration to Letter XXX, 3 April 1776.

'Now as it appears that this bird, which is so well known to practice incubation, is formed in a similar manner with cuckoos, Monsieur Herissant's conjecture, that cuckoos are incapable of incubation from the disposition of their intestines, seems to fall to the ground; and we are still at a loss for the cause of that strange and singular peculiarity of the instance of the *Cuculus canorus*.'

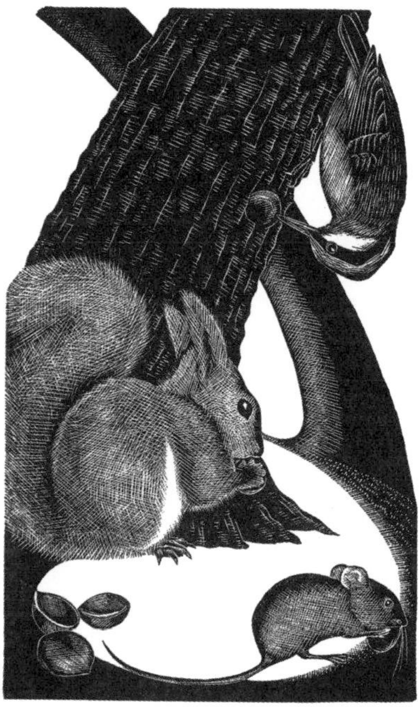

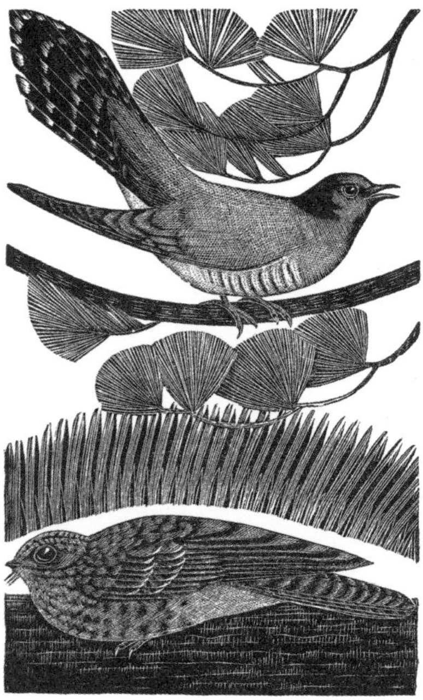

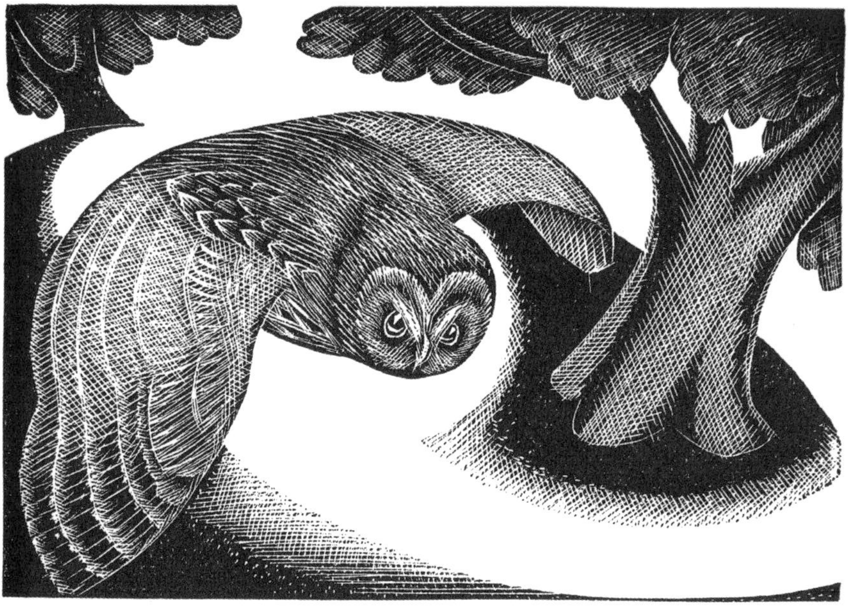

FIG.128
Owl. Illustration to Letter XLIII, Selborne, 9 September 1778.

'The language of birds is very ancient, and, like other modes of speech, very elliptical; little is said, but much is meant and understood...Owls have very expressive notes; they hoot in a fine vocal sound, much resembling the *vox humana*, and reducible by a pitch-pipe to a musical key.'

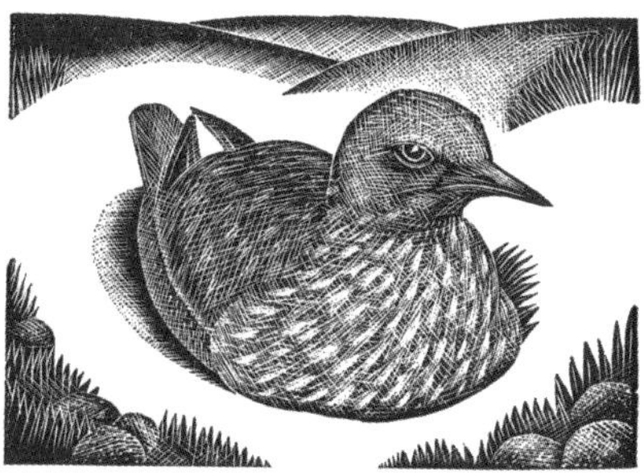

FIG.129
Stone Curlew. Illustration to Letter XXXIII, 26 November 1770.

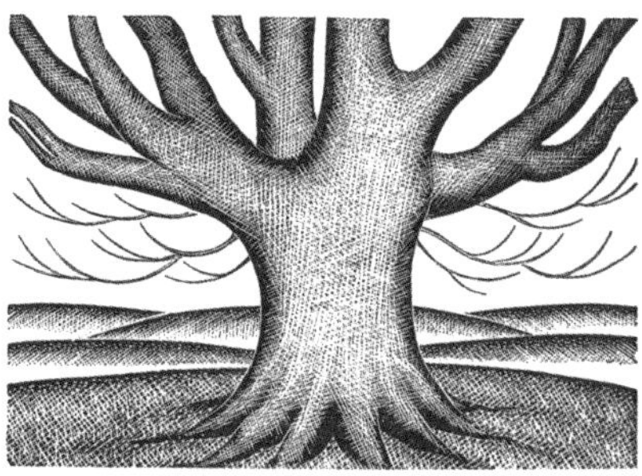

FIG.130
Beech Tree. Illustration to Letter I to Thomas Pennant.

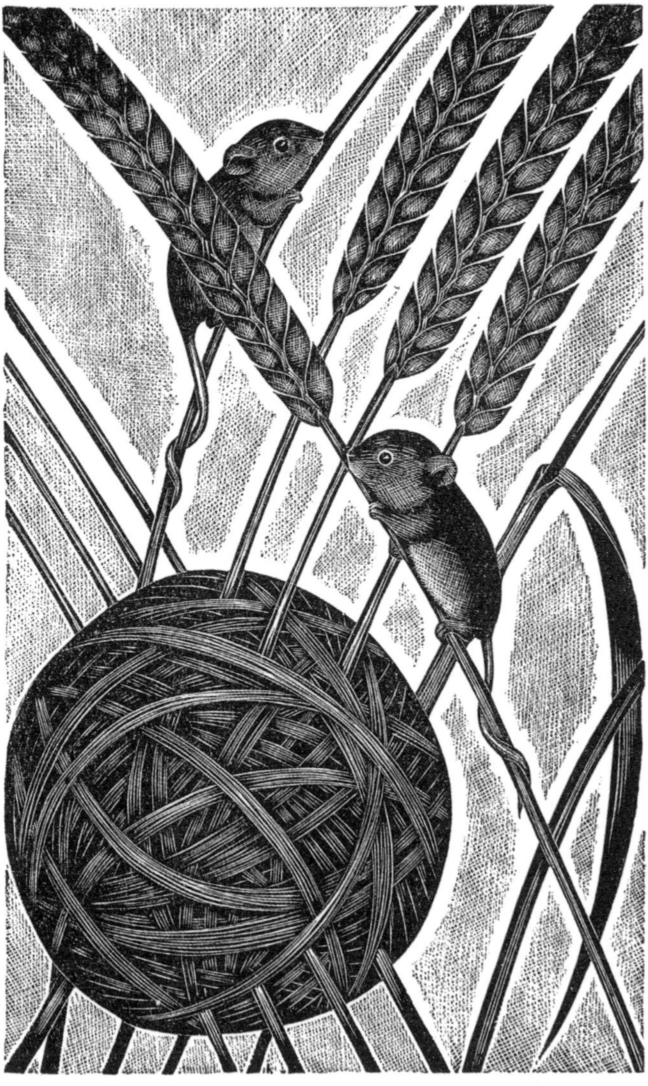

FIG.131
Harvest Mouse and Nest. Illustration to Letter XIII, Selborne, 22 January 1768.

'As to the small mice, I have farther to remark, that though they hang their nests for breeding up amidst the straws of the standing corn, above the ground; yet I find that, in the winter, they burrow deep in the earth, and make warm beds of grass: but their grand rendezvous seem to be in cornricks, into which they are carried at harvest. A neighbour housed an oat-rick lately, under the thatch of which were assembled nearly a thousand, most of which were taken, and some I saw. I measured them; and found that, from nose to tail, they were just two inches and a quarter, and their tales just two inches long. Two of them, in a scale, weighed down just one copper half-penny, which is about the third of an ounce avoirdupois; so that I suppose they are the smallest quadruped in this island.'

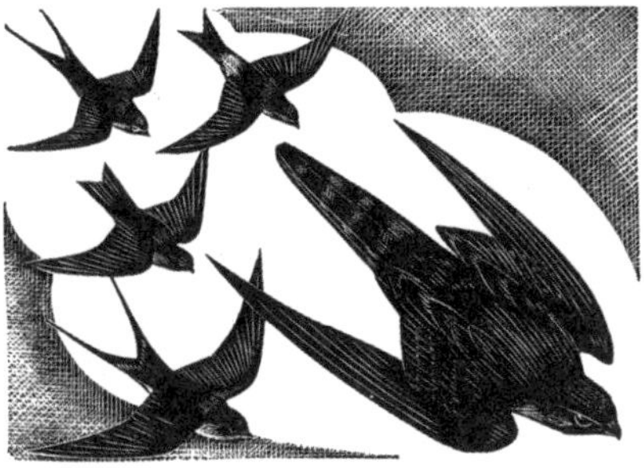

FIG.132
Swallows Mobbing a Hawk. Illustration to Letter XVIII, Selborne, 29 January 1774.

'The swallow, probably the male bird, is the *excubitor* to house martins, and other little birds, announcing the approach of birds of prey. For as soon as the hawk appears, with a shrill alarming note he calls all the swallows and martins about him; who pursue in a body, and buffer and strike their enemy until they have driven him from the village, darting down from above on his back, and rising in a perpendicular line in perfect security.'

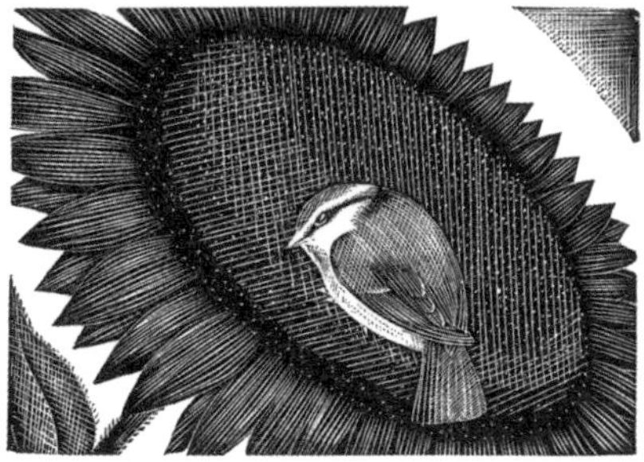

FIG.133
Blue Tit. Illustration to Letter XLI.

'The blue titmouse, or nun, is a great frequenter of houses, and a general devourer. Besides insects, it is very fond of flesh, for it frequently picks bones on dunghills; it is a vast admirer of suet and haunts butcher's shops… It will also pick holes in apples left on the ground, and be well entertained with the seeds on the head of a sunflower.'

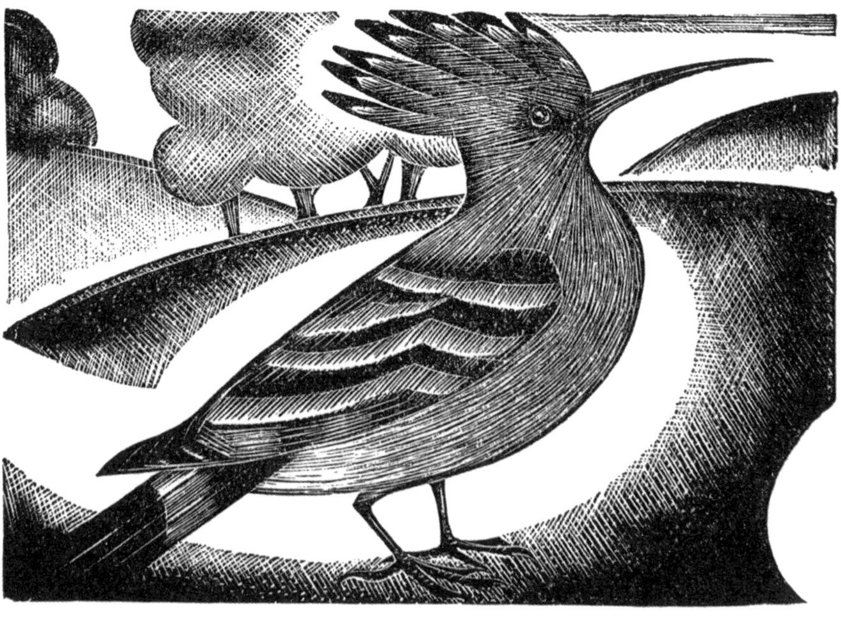

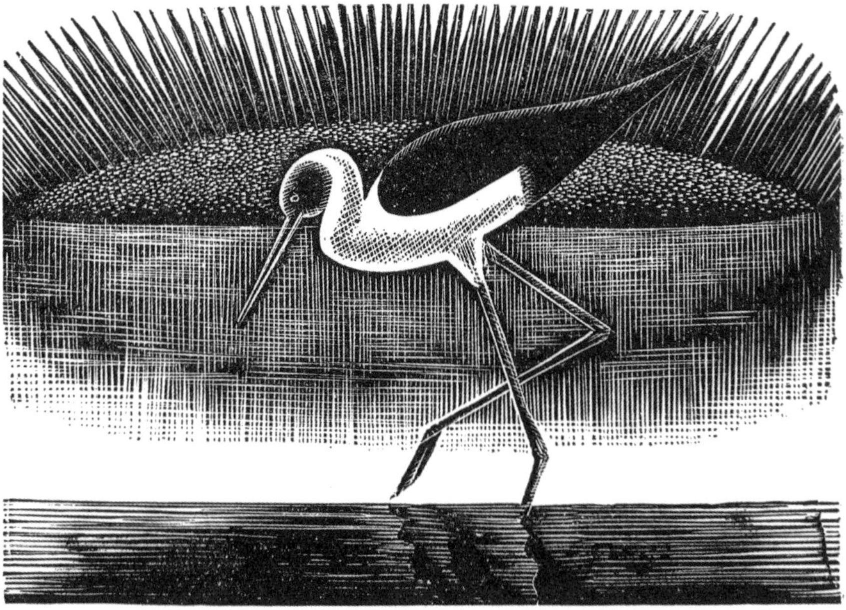

FIG.135
Stilt. Illustration to Letter XLIX, Selborne, 7 May 1779.

'It must be matter of great curiosity to see the stilt plover move; to observe how it can wield such a length of lever with such feeble muscles as the thighs seem to be furnished with. At best one should expect it to be but a bad walker: but what adds to the wonder is, that is has no back toe. Now without that steady prop to support its steps it must be liable, in speculation, to perpetual vacillations, and seldom able to preserve the true centre of gravity.'

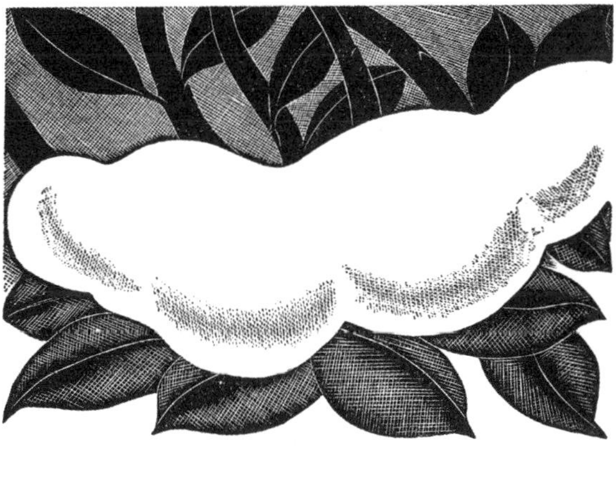

FIG.134 (opposite, below)
Hoopoe.

FIG.136
Evergreens in Snow.

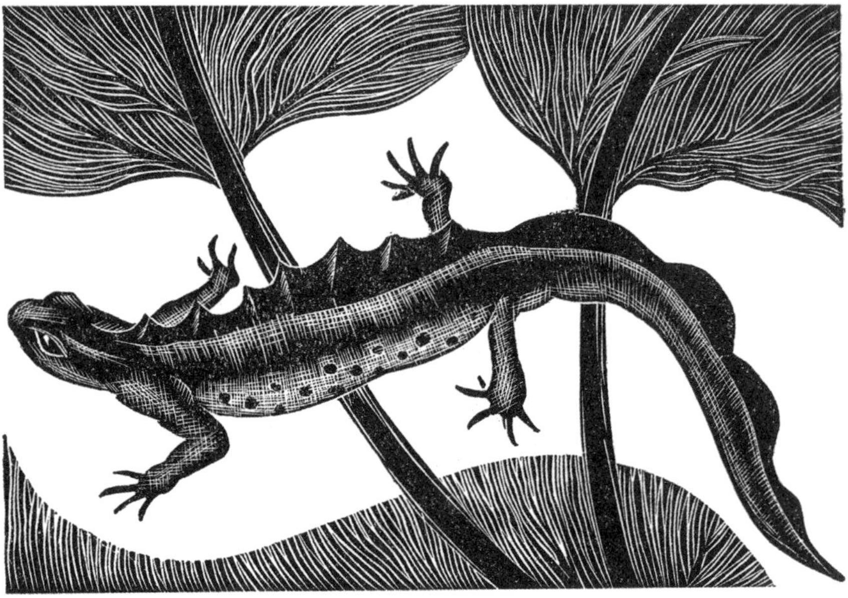

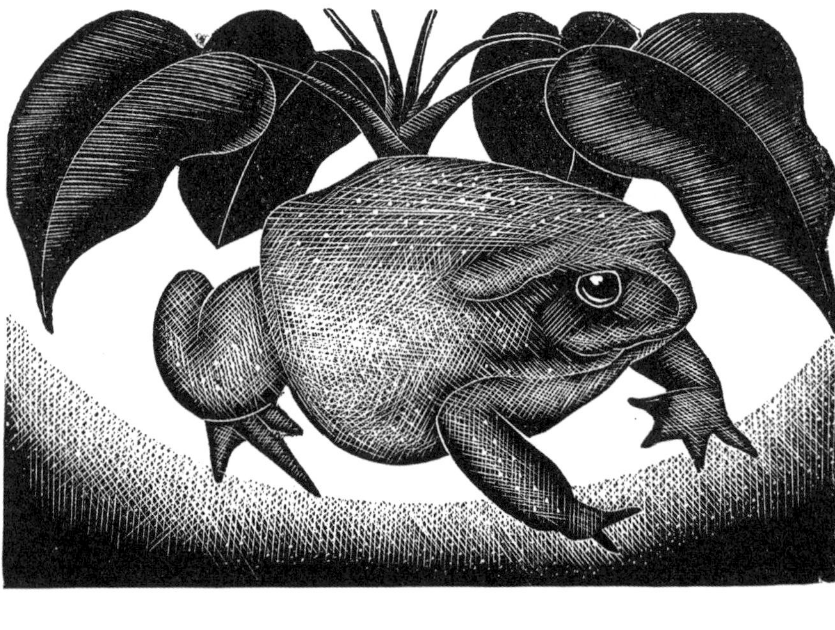

FIG.137 (opposite, above)
Newt. Illustration to Letter XX, Selborne, 8 October 1768.

'I must not omit to tell you (as I have been so lately on the study of reptiles) that my people, every now and then of late, draw up with a bucket of water from my well, which is sixty-three feet deep, a large black warty lizard with a fin-tale and yellow belly. How they first came down at that depth, and how they were ever to have got out thence without help, is more than I am able to say.'

FIG.138 (opposite, below)
Toad. Illustration to Letter XVII, Selborne, 18 June 1768.

'The method in which toads procreate and bring forth seems to be very much in the dark. Some authors say that they are viviparous: and yet Ray classes them among his oviparous animals, and is silent with regard to the manner of their bringing forth. Perhaps they may be έσω μεν ωοτόκου, έξω δε ζωοτόκου ('inside laying eggs, outside laying eggs') as is known to be the case with the viper.'

FIG.139
Tortoise. Illustration to Letter XIII, 12 April 1772.

'On the 1st November I remarked that the old tortoise, formerly mentioned, began first to dig the ground in order to the forming its hybernaculum, which it had fixed on just beside a great tuft of hepaticas. It scrapes out the ground with its forefoot, and throws it up over its back with its hind; but the motion of its legs is ridiculously slow, little exceeding the hour-hand of a clock; and suitable to the composure of an animal said to be a whole month in performing one feat of copulation. Nothing can be more assiduous than this creature night and day in scooping the earth, and forcing its great body into the cavity; but, as the noons of that season proved unusually warm and sunny, it was continually interrupted, and called forth by the heat by the middle of the day; and though I continued there until the 13th November, yet the work remained unfinished.'

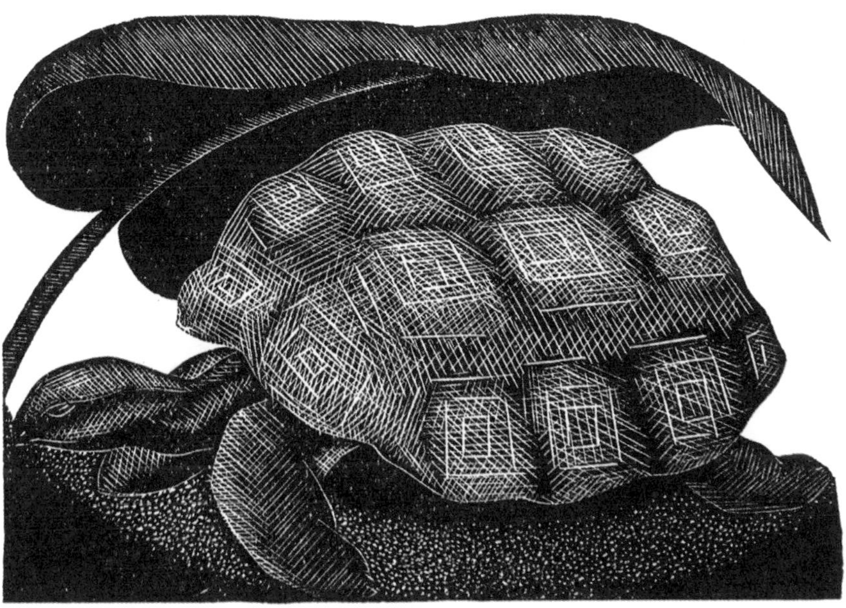

JOHN NASH R.A. (1893–1977)

Gilbert White, *The Natural History of Selborne with Drawings by John Nash,* John Lewis, ed., London: Lutterworth Press, London, 1951. Black and white line letterpress

Gilbert White, *The Natural History of Selborne,* Ipswich: Limited Editions Club, 1972. Colour lithographs

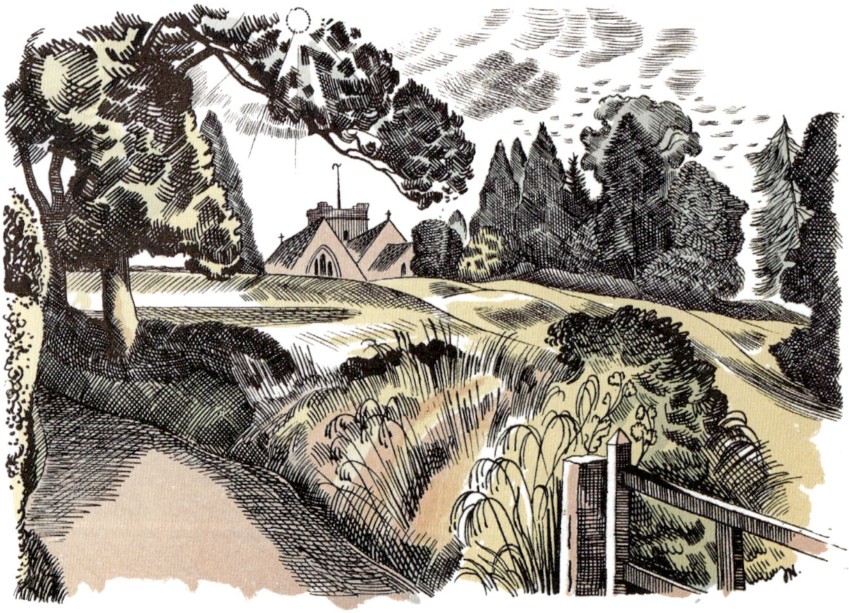

FIG.140
Selborne Church from the Lithe, 1951.

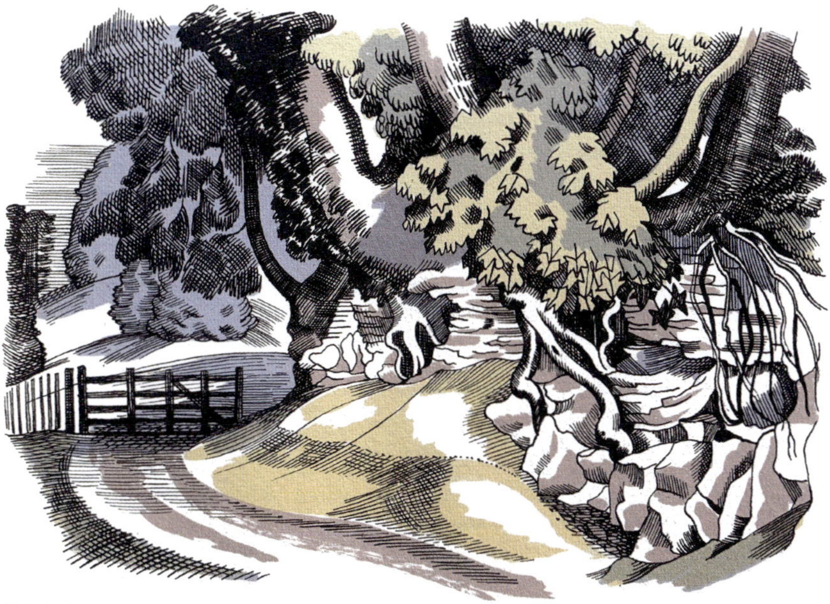

FIG.141
Rocky Lane near the Lithe, 1951/ 1972.

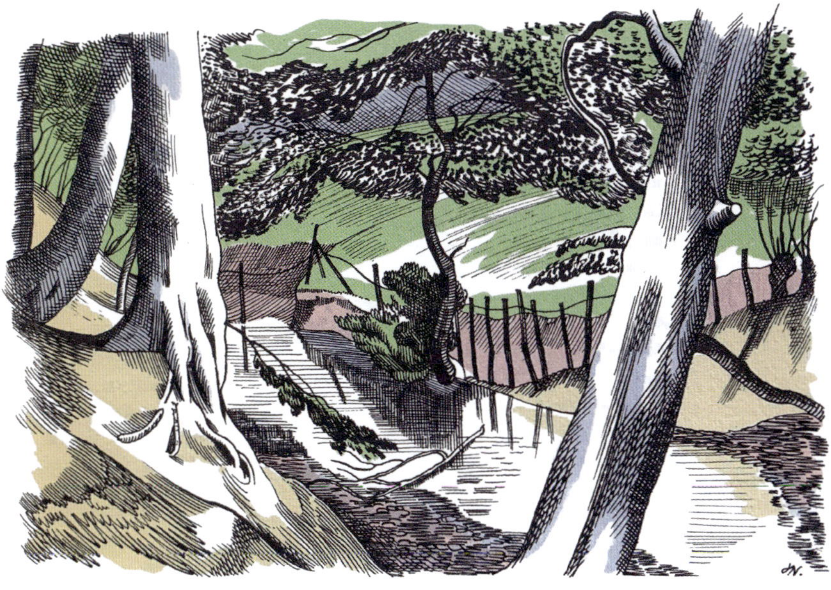

FIG.142
Stream in the Lithe, 1951/ 1972.

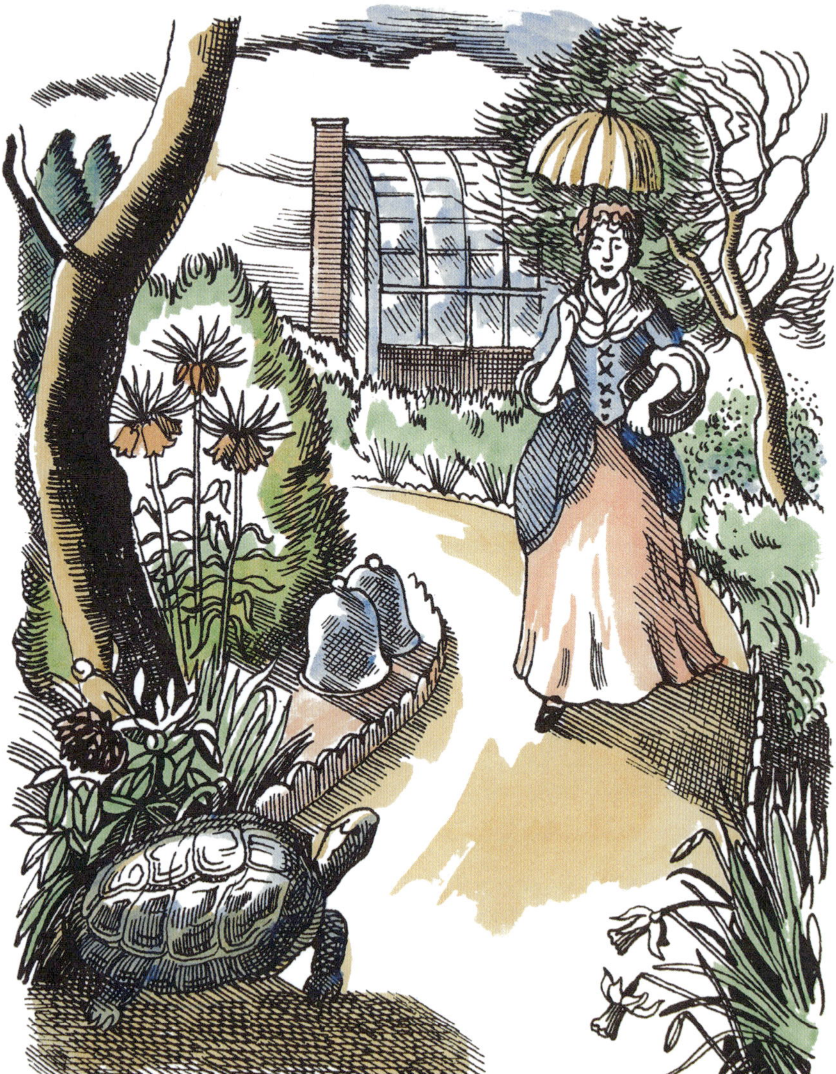

FIG.143
'It hobbles towards its benefactress', 1951/1972. Illustration to Letter XIII to the Hon. Daines Barrington, 12 April 1772.

'I was much taken with its sagacity in discerning those that do it kind offices: for, as soon as the good old lady comes in sight who has waited on it for more than thirty years, it hobbles towards its benefactress with awkward alacrity; but remain inattentive to strangers. Thus not only "the ox knoweth his owner, and the ass his master's crib,' but the most abject reptile and torpid of beings distinguishes the hand that feeds it, and is touched with the feelings of gratitude!'

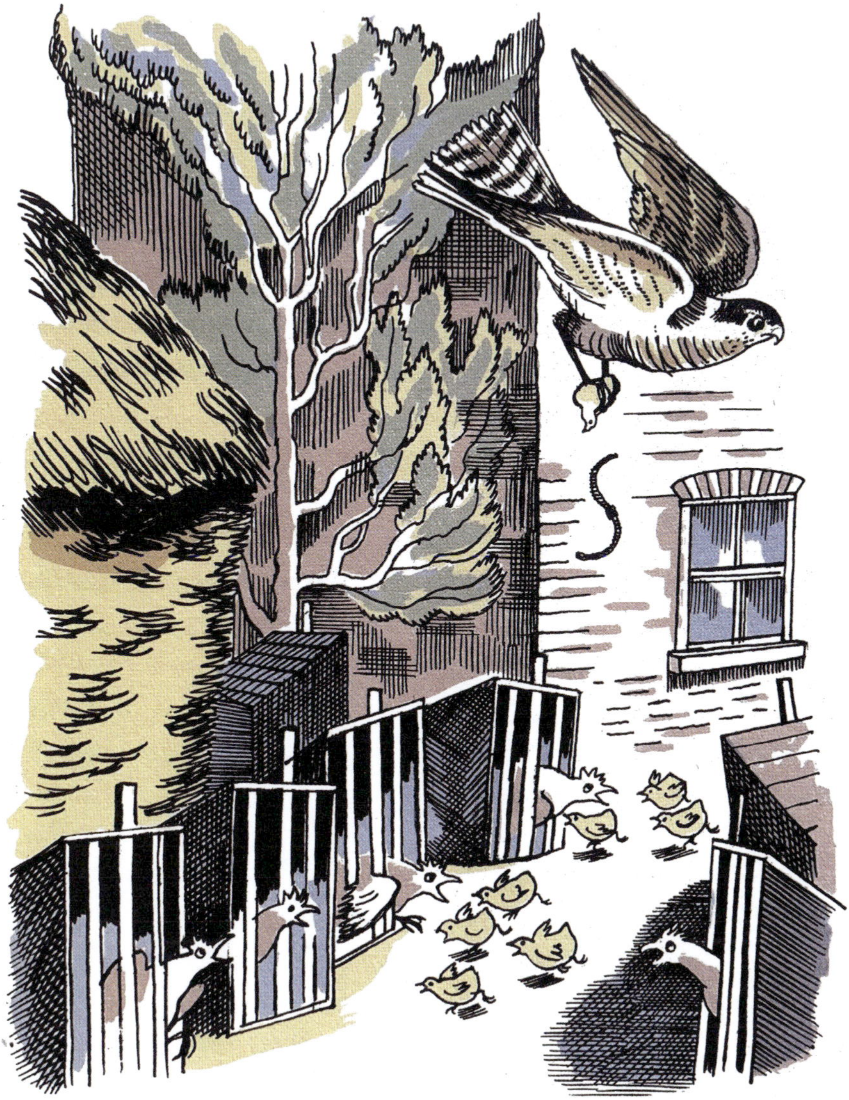

FIG.144
'A terror to all ... that had chickens',
1951/1972. Illustration to letter XLIII to Thomas Pennant.

'About the tenth of July in the same summer a pair of sparrow-hawks bred in an old crow's nest on a low beech in the same hanger; and as their brood, which was numerous, began to grow up, became so daring and ravenous, that they were a terror to all the dames in the village that had chickens or ducklings under their care.'

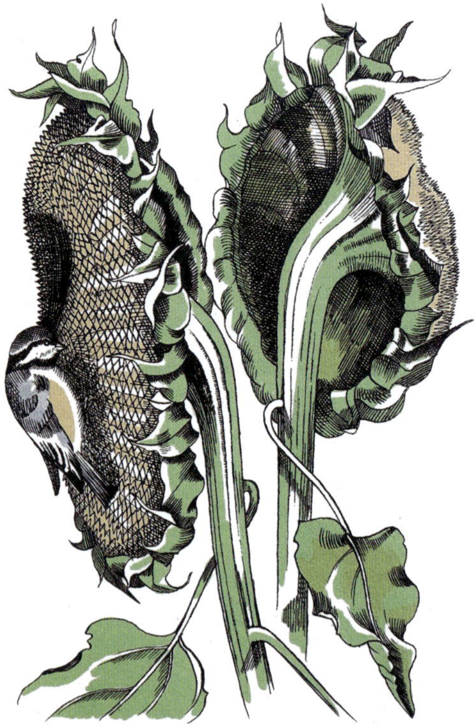

FIG.145
'Well entertained with a sunflower', 1951/1972. Illustration to Letter XLI to Thomas Pennant.

FIG.146
Stinking Hellebore, 1951/1972.

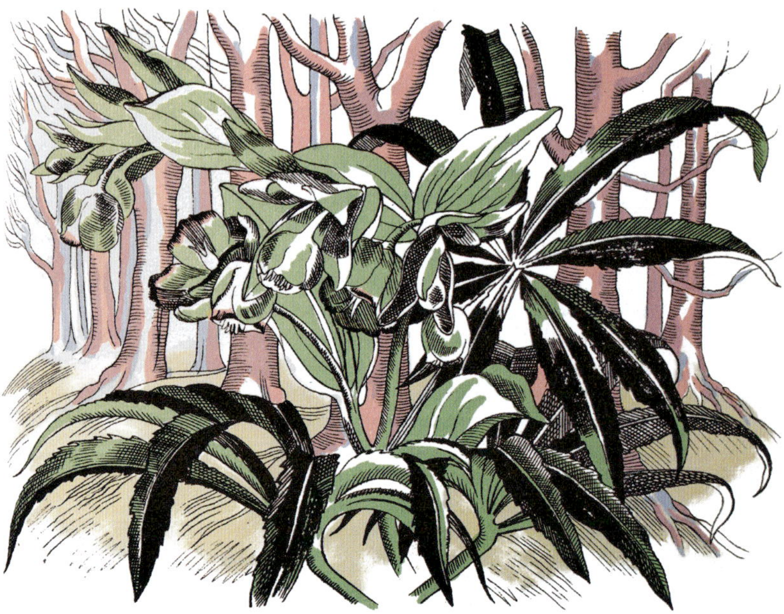

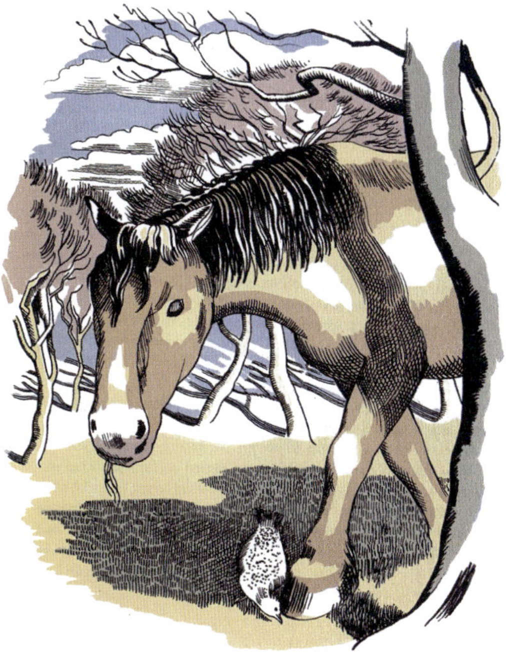

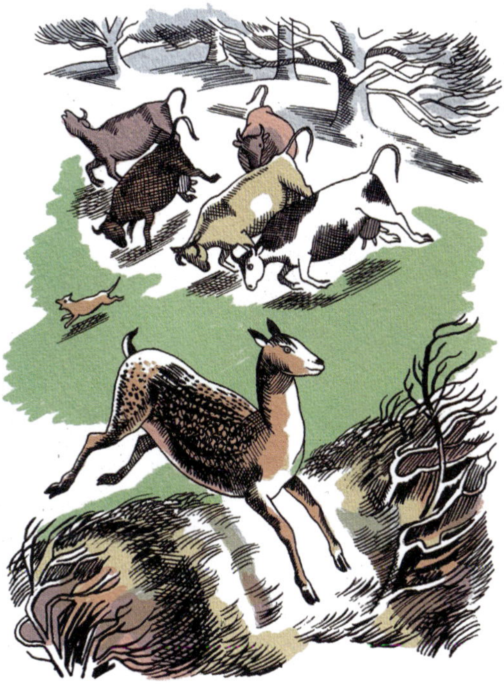

FIG.147
'Each seemed to console the other', 1951/1972. Illustration to Letter XXIV to the Hon. Daines Barrington, 15 August 1775.

FIG.148
'Drive the assailants right out of the pasture', 1951/1972. Illustration to Letter XXIV to the Hon. Daines Barrington, 17 August 1775.

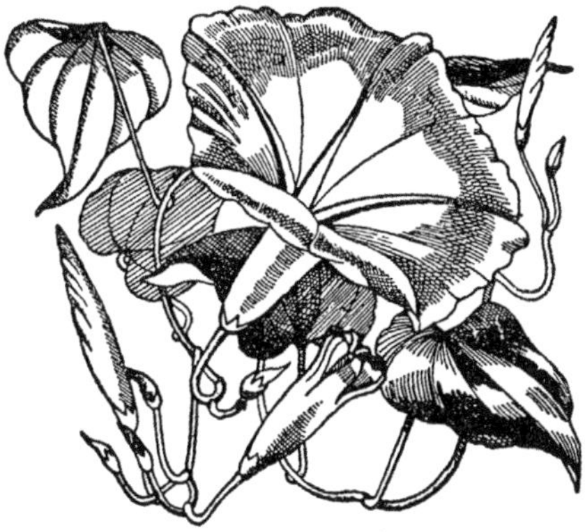

FIG.149
Frontispiece, 1972.

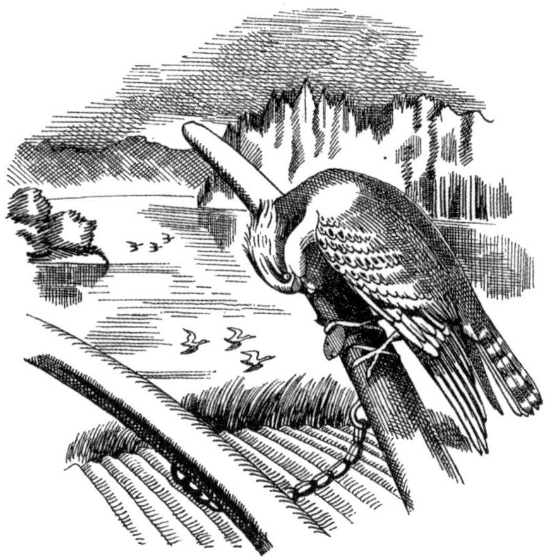

FIG.150
The Osprey, 1951.

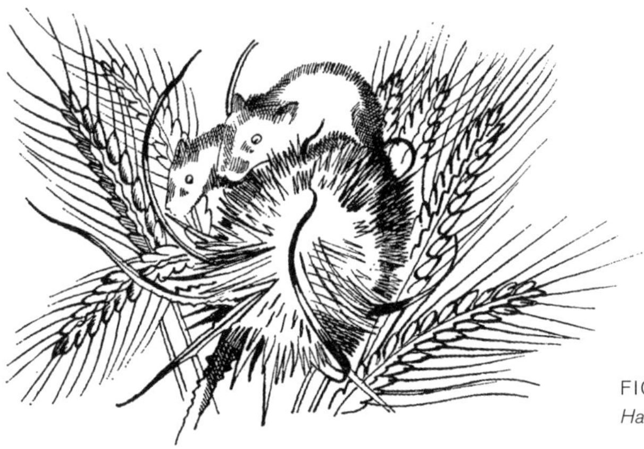

FIG.151
Harvest Mice, 1951.

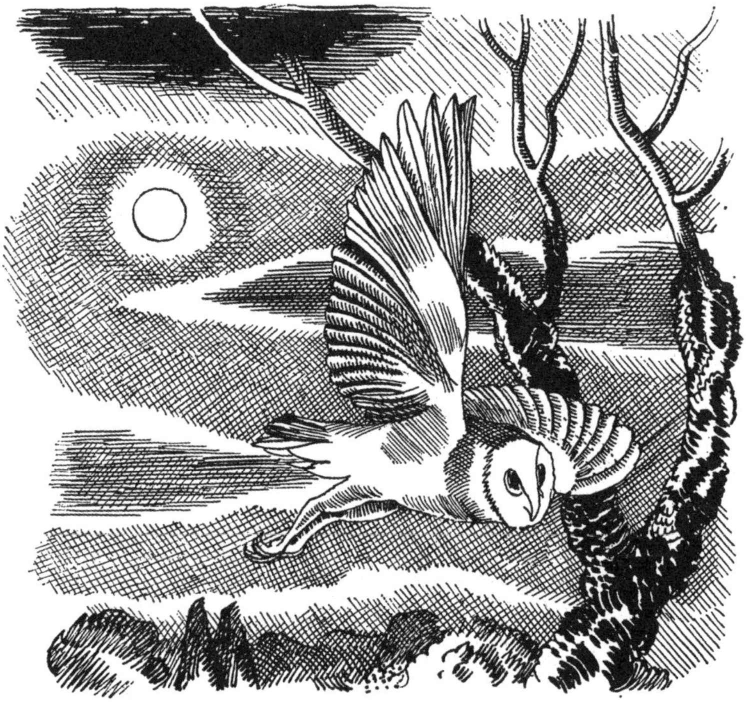

FIG.152
A Young Barn Owl, 1951.

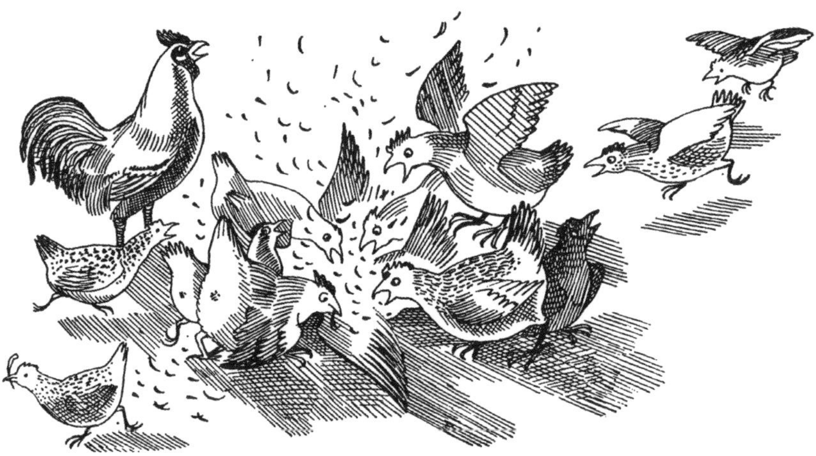

FIG.153
Revenge! 1951.

FIG.154
'The common snake in pursuit of frogs', 1951. Illustration to Letter XVII to Thomas Pennant.

'Country people talk much of a water-snake, but, I am pretty sure, without any reason; for the common snake (*coluber natrix*) delights much to sport in the water, perhaps with a view to procure frogs and other food.' Letter XVII, Selborne, 18 June 1768.

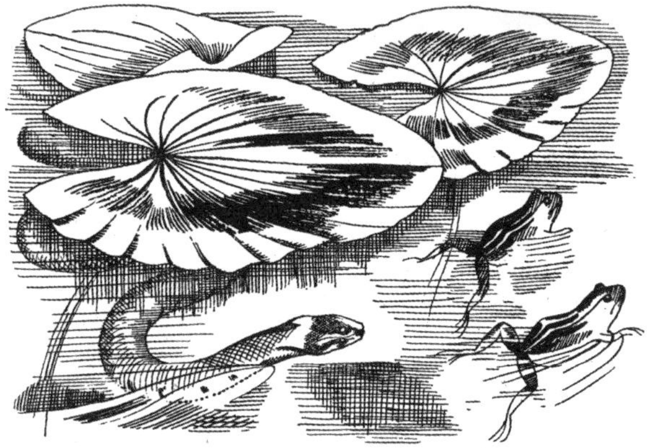

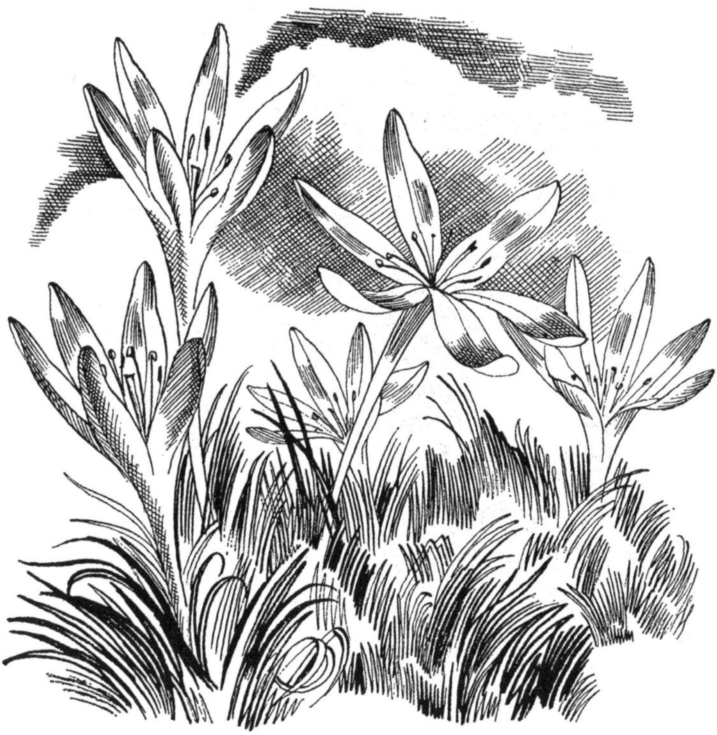

FIG.155
Autumn Crocus, 1951.

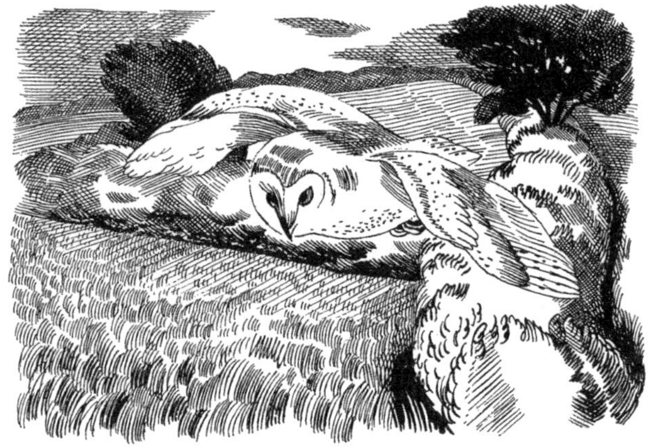

FIG.156
'About an hour before sunset they sally forth', 1951.

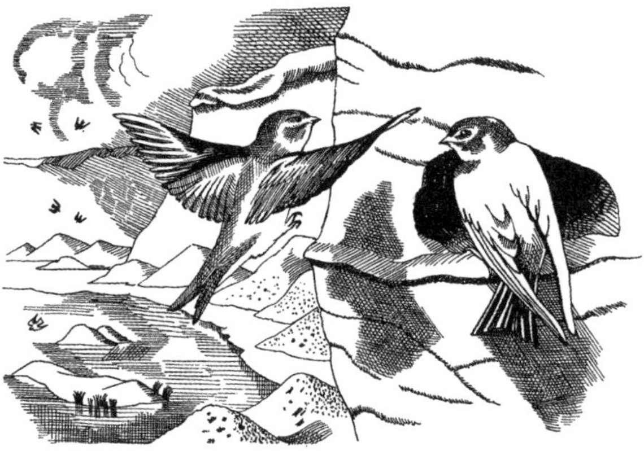

FIG.157
Sand-martins, 1951. Illustration to Letter XX to the Hon. Daines Barrington, 26 February 1774.

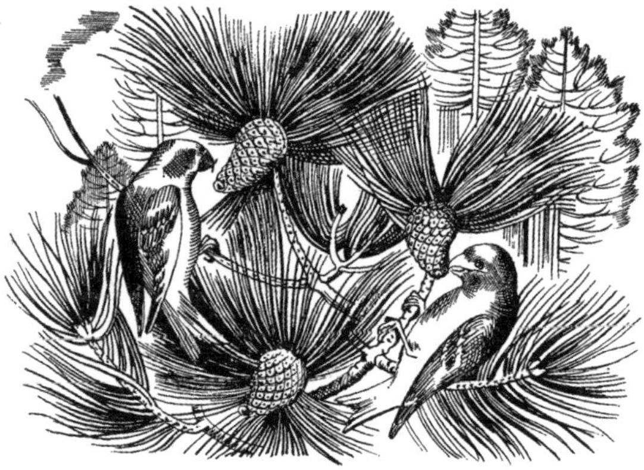

FIG.158
'A flight of cross-bills comes to the pine-grove', 1951. Illustration to Letter XVII to the Hon. Daines Barrington, 9 December 1773.

JOHN PIPER CH (1903–92)

1962: pen, ink and wash drawings
Gilbert White, *The Natural History of Selborne, edited with an Introduction and Notes by W.S. Scott and Drawings by John Piper*, London: Folio Society, 1962

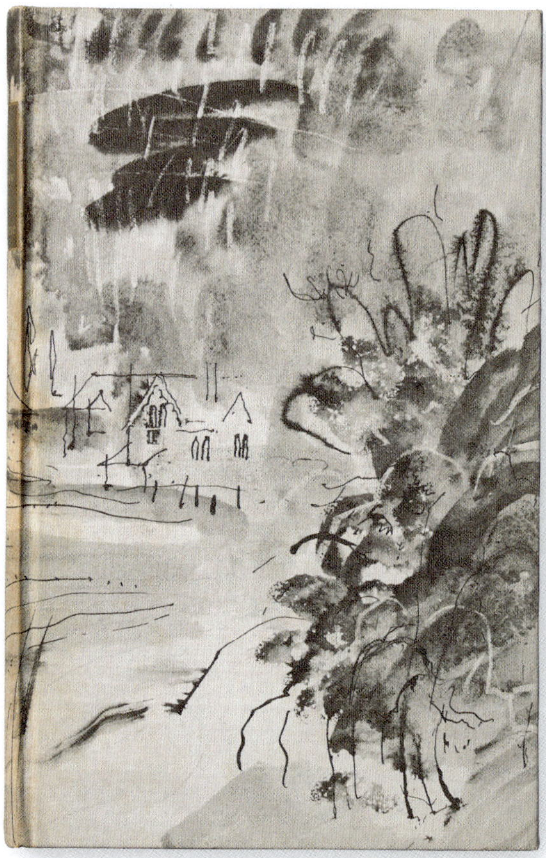

FIG.159
Cover design.

FIG.160 (opposite)
From the Hanger.
Pen, ink and wash drawing, reproduced in collotype.

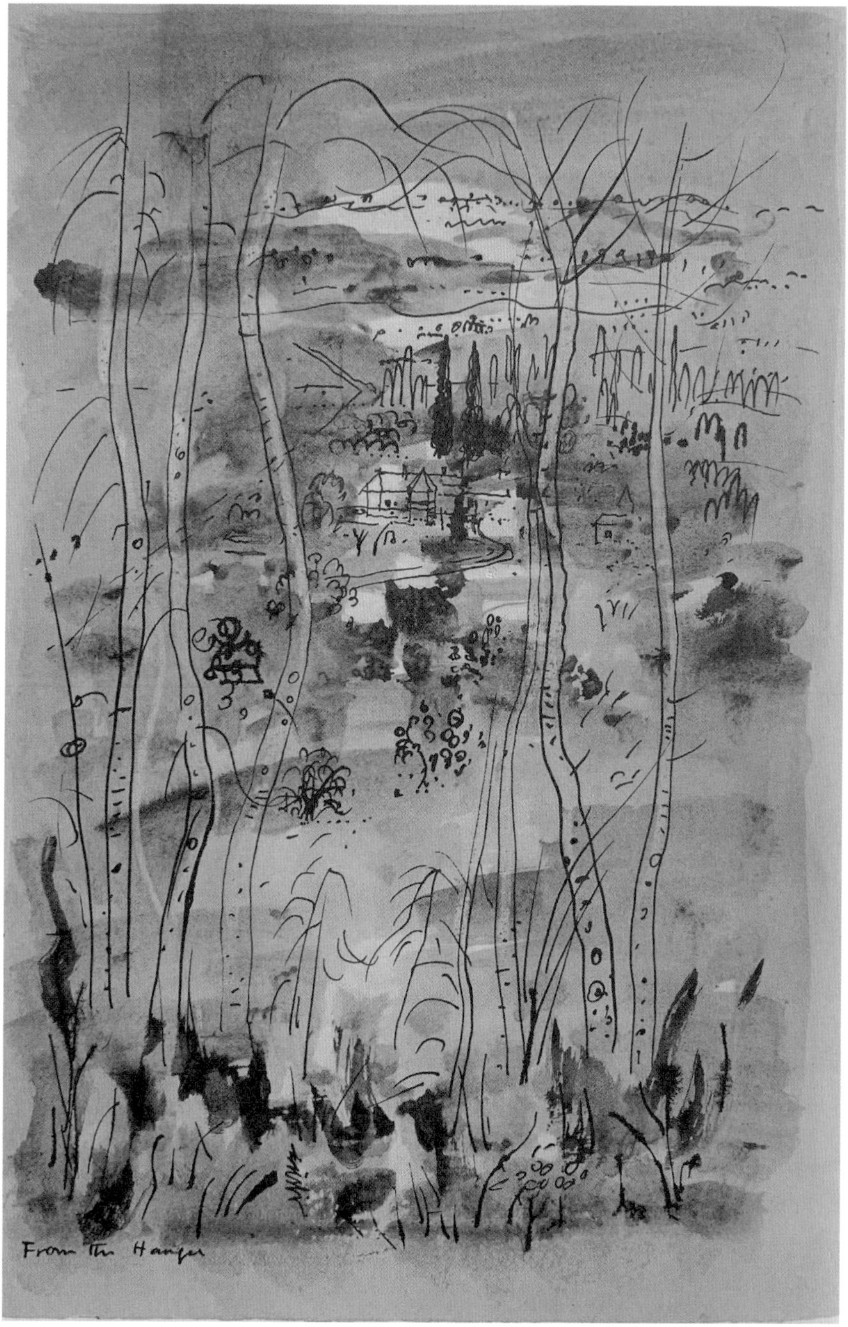
From the Hanger

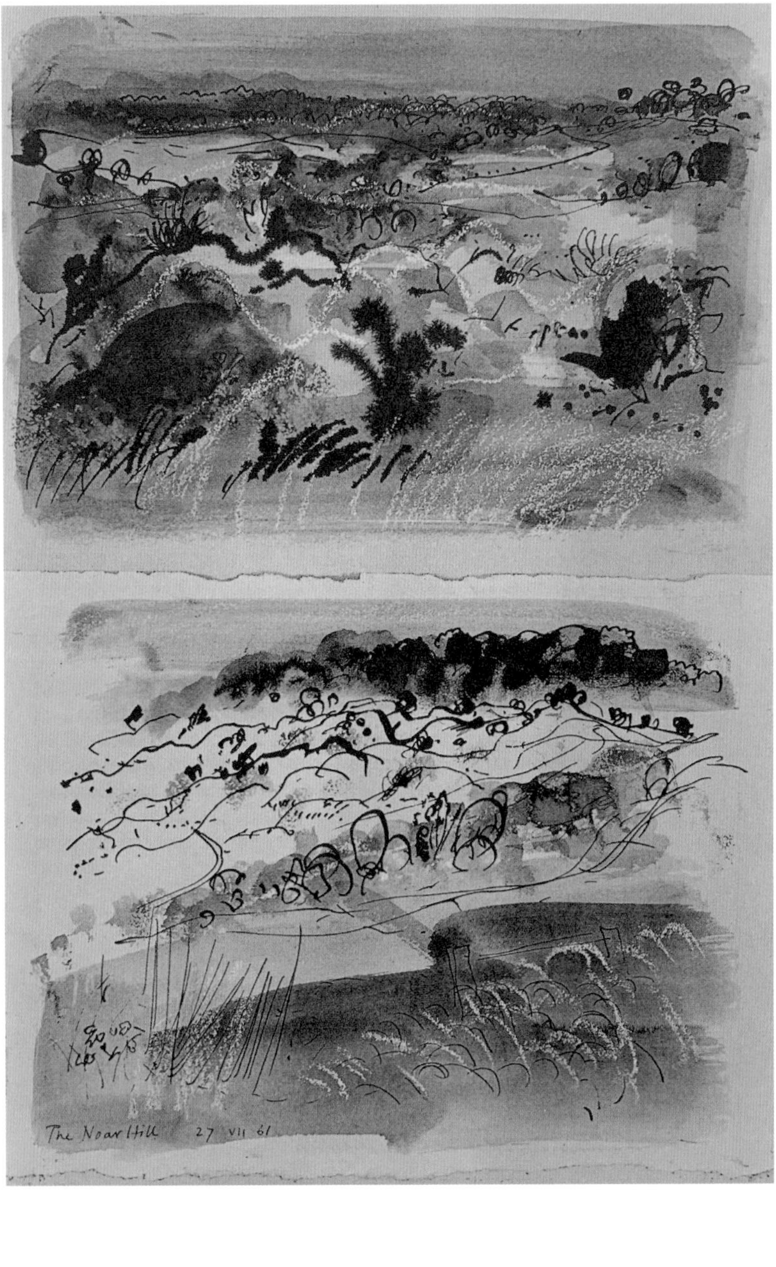

FIG.161
The Noar Hill. Pen, ink and wash drawing, reproduced in collotype.

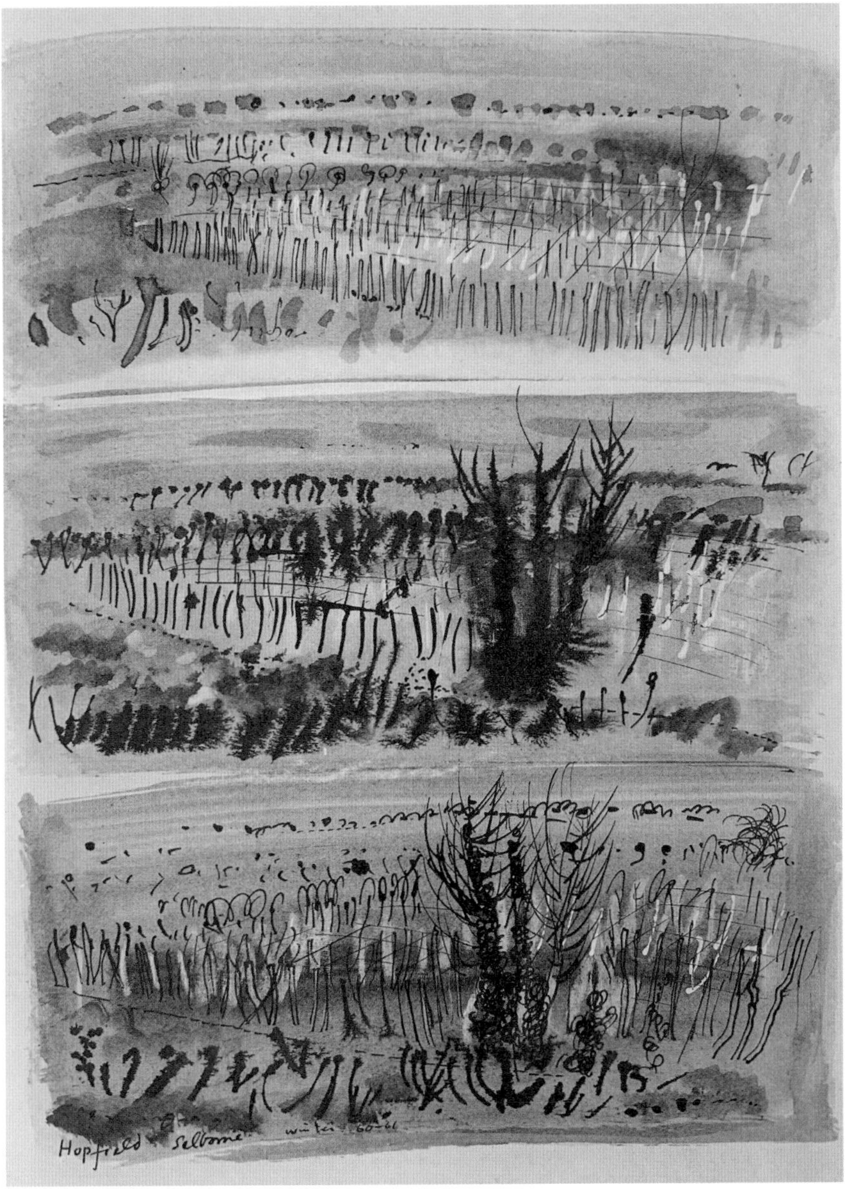

FIG.162
Hopfield Selbourne. Pen, ink and wash drawing, reproduced in collotype.

CHRISTOPHER WORMELL (b.1955)

1994: wood-engravings

Gilbert White, *The Natural History of Selborne,* introduced by Ian Niall, London: Folio Society, 1994

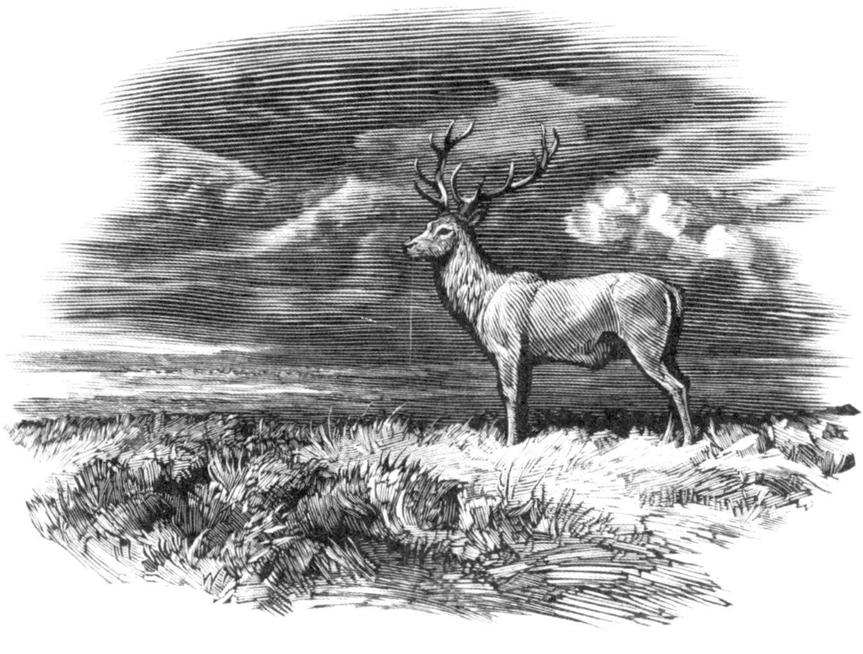

FIG.163
Stag.

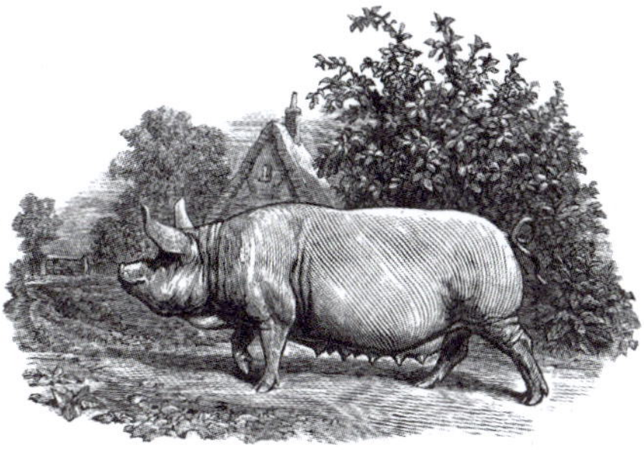

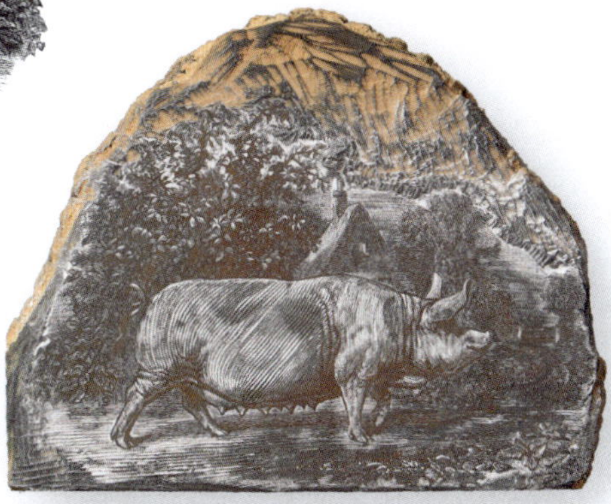

FIG.164a–b
Sow. Wood-engraving on paper and engraved block.

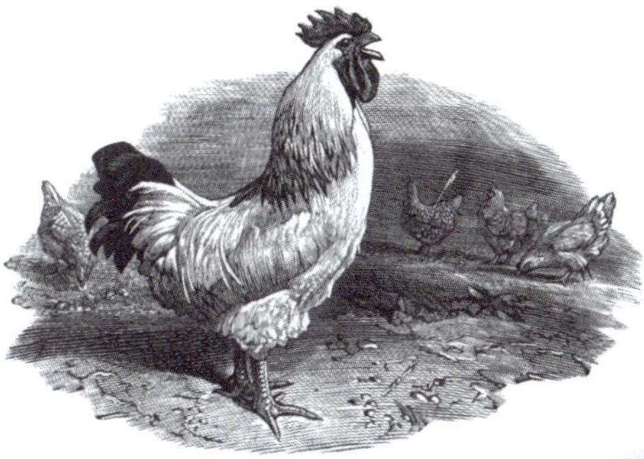

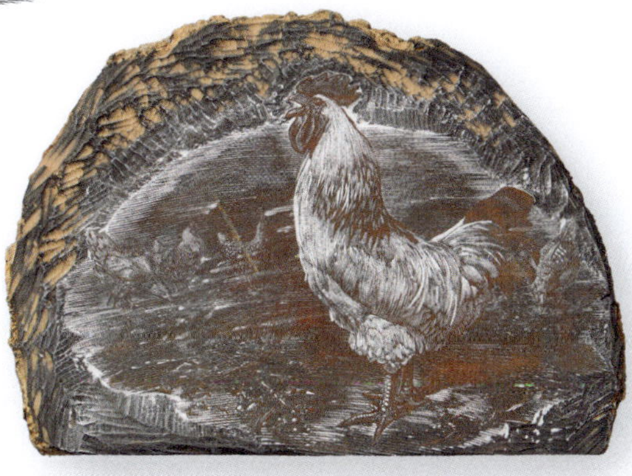

FIG.165a–b
Cockerel. Wood-engraving on paper and engraved block.

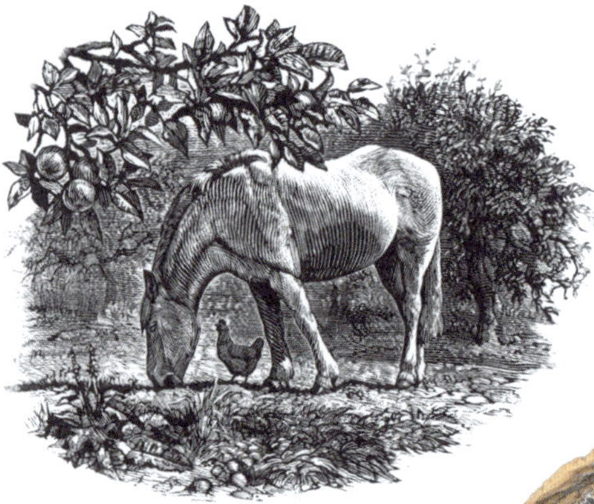

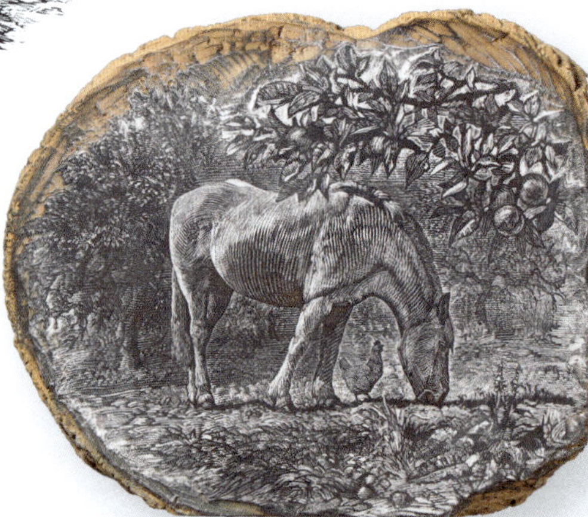

FIG.166a–b
Horse and Hen.
Wood-engraving on paper
and engraved block.

FIG.167
Barn Owl.

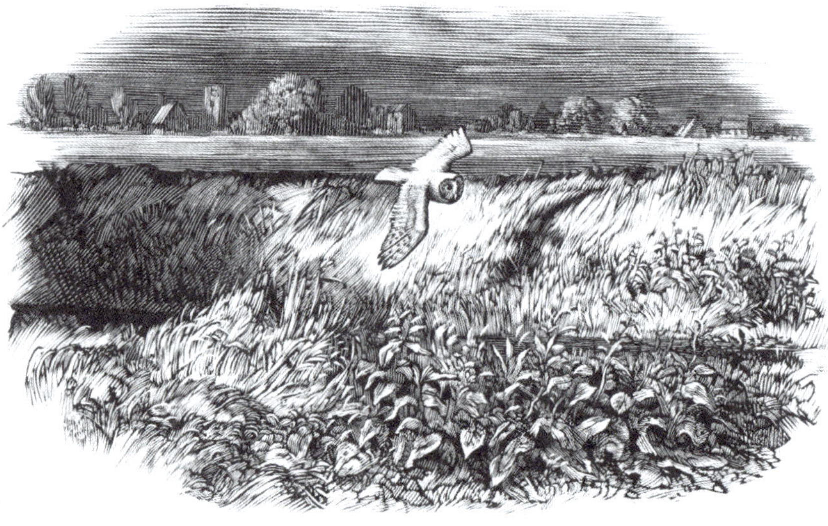

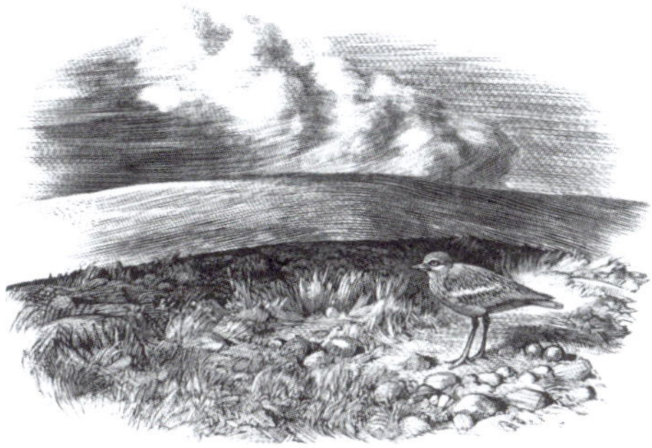

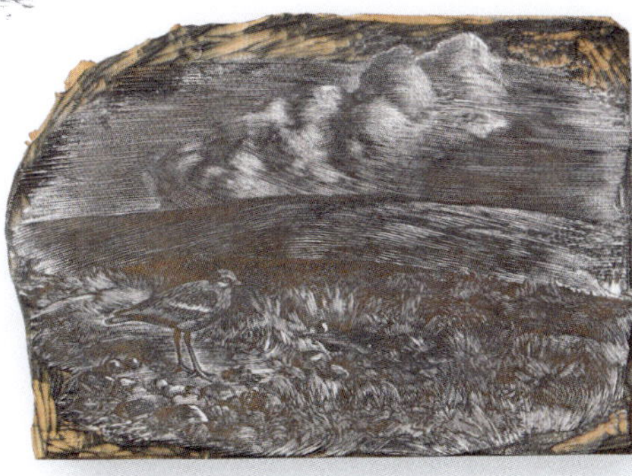

FIG.168a–b
Stone Curlew. Wood-engraving on paper and engraved block.

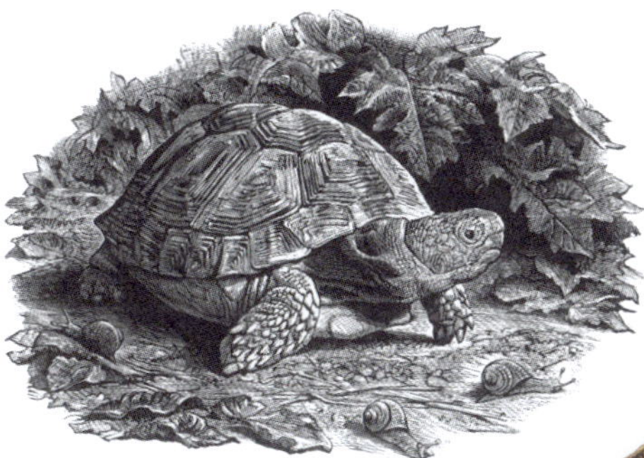

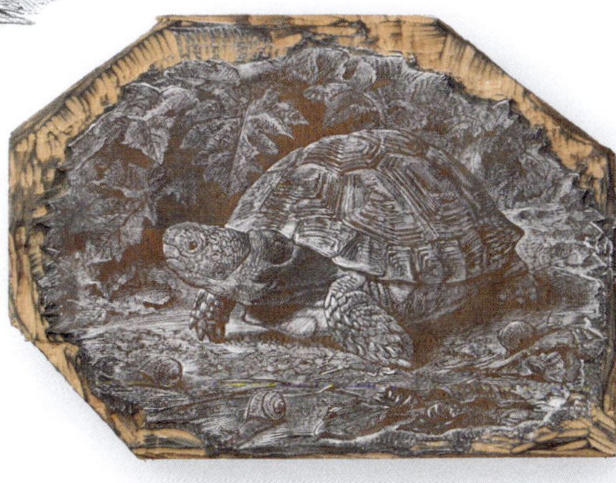

FIG.169a–b
Tortoise. Wood-engraving on paper and engraved block.

FREDERICK MARNS (1933–2013)

1977: pencil drawings
Gilbert White, *The Natural History of Selborne*, with preface by Lord Selborne and an introduction by Sir David Attenborough, London: Shepheard-Walwyn, 1977

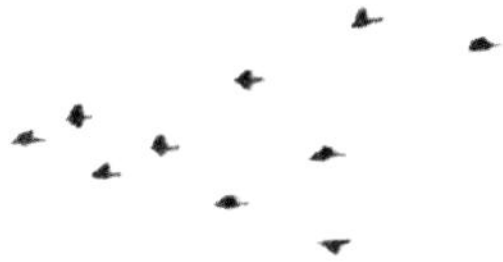

FIG. 170
Dorton Cottage.

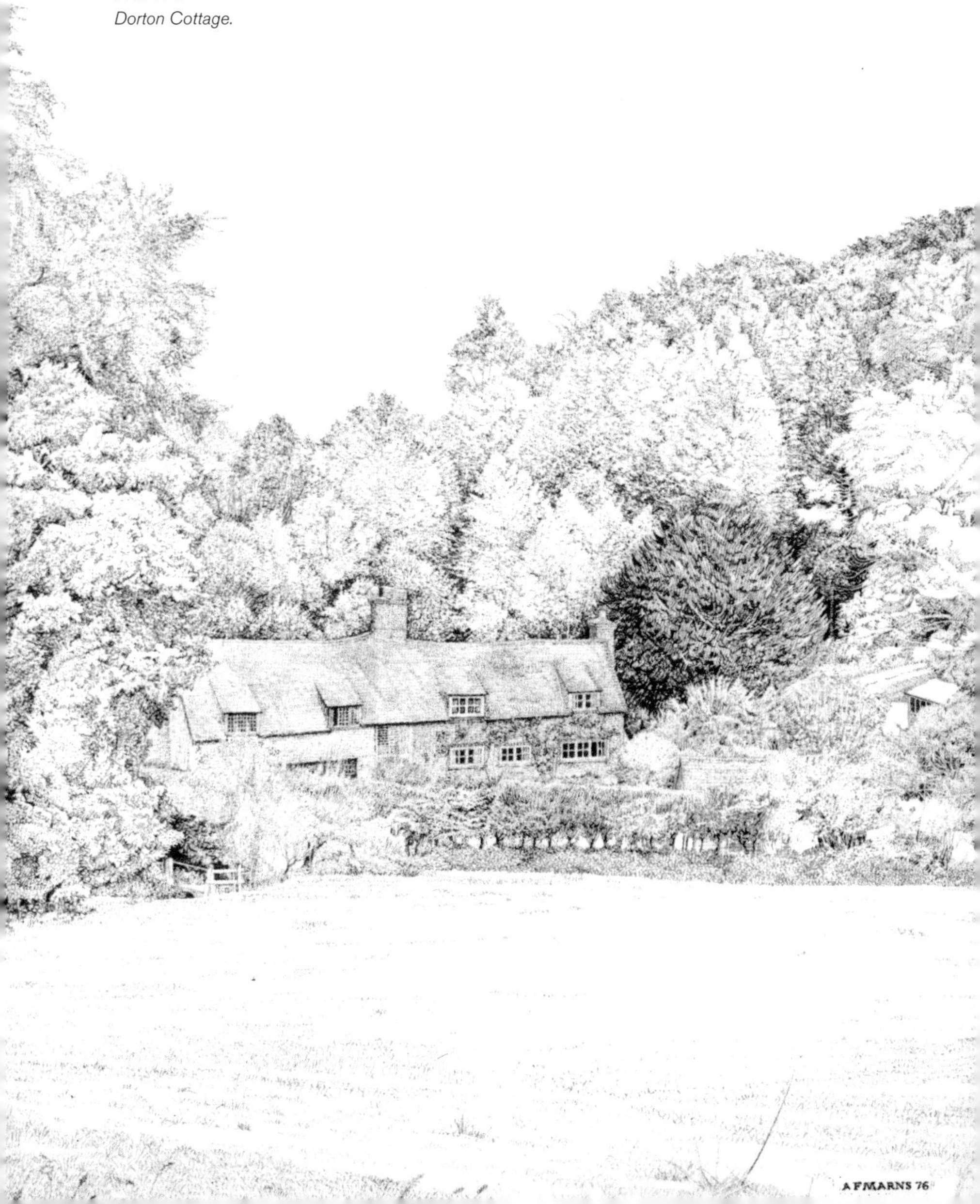

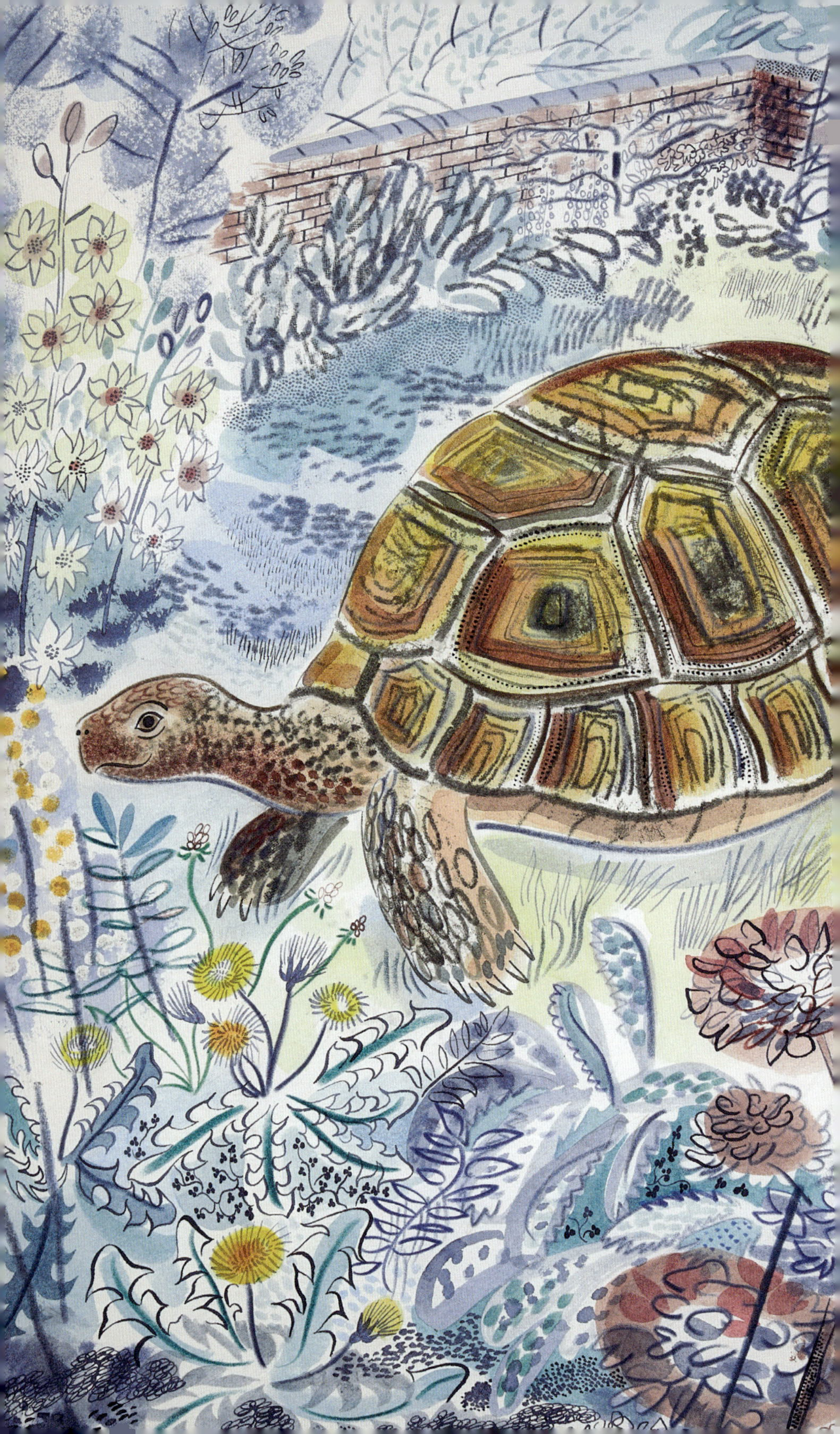

5.

CONTEMPORARY RESPONSES TO GILBERT WHITE

Simon Martin

Over 230 years on from its first publication, Gilbert White's *Natural History of Selborne* engages readers and artists today. Part of its continuing appeal is undoubtedly the way in which each generation has found a path into the book that has relevance to their times, and how successive artists have brought to it their own form of expression. It was during my first visit to the Gilbert White & the Oates Collections in spring 2019 that I first came under the charm of Selborne. On noticing that the tercentenary of the naturalist's birth was approaching in the following year, it struck me that there was an opportunity to create a timely exhibition of works by the various Modern British artists who had illustrated the book. Returning home, I found my copy of the *Natural History*, the Folio Society edition with illustrations by John Piper. In the subsequent months, I started to make lists of all the artists who had illustrated the book, and gradually piles of copies of the *Natural History* started building up around me. The result was the exhibition *Drawn to Nature: Gilbert White and the Artists* at Pallant House Gallery in 2020.[1]

FIG.171
Detail from *Tortoise*, 2020 (see fig.188).

One might assume that the artistic response to Gilbert White was well-charted territory, but seemingly not, aside from a small exhibition in 1989 to mark the bicentenary of the publication at the Manchester Polytechnic Library.[2] Inevitably, this is a narrative that oscillates between the particular and the general, between the specificity of White's narrative and the wider story of natural history illustration and modern artists' books. Rather than simply presenting a series of open tomes, where possible the exhibition included the artists' original impressions or drawings. These have become part of the unique and unfolding history of how one book can have inspired so many artists. Yet it was important that the exhibition also considered the relevance of Gilbert White's writings on nature to artists now. With this in mind, I approached eleven contemporary printmakers and illustrators whom I felt had an affinity to the themes of nature in the book and to the legacy of the artists who had tackled it in the past, inviting them each to create their own response to the *Natural History.* What has emerged from this commission is an ongoing, thoughtful and self-conscious history – and evidence of the continuous enthusiasm amongst artists for White's writings and evocative images. As Mark Hearld, designer of this book's cover, observed, 'the text is so visual and White is so particular. Each letter is a perfect distillation of a single moment. You can imagine White writing it every time.'[3]

Clive Hicks-Jenkins (b.1951), a Welsh artist who has illustrated numerous books on themes connected to animal life, such as *Equus* (2013) and *The Mare's* Tale (2001), has described the *Natural History*, together with *Kilvert's Diary 1870–79* by another clergyman naturalist Francis Kilvert (1840–79) as 'the nature reference books of my youth'. He noted the impact of White and Bewick on his own pencil-drawn subjects drawn from natural history including an owl, goat, ox, swan, hare (fig.180) and cockerel for Marly Youman's novel *Charis in the World of Wonders.*[4] Most of the contemporary artists selected have illustrated books on themes relating to nature, often aimed at introducing children and young people to ideas of biodiversity. Emily Sutton (b.1983) has illustrated several science and nature books, such as *Lots: The Diversity of Life* (2014) and *Tiny Creatures: The Invisible World of Microbes* (2014). Her characterful watercolour *White Owl* (fig.189) has the full moon doubling up as the letter 'O', echoing a series of alphabet prints that Sutton created including

'O for Owl' with an aesthetic inspired by Old German educational charts and a nod to the owls that flew around her parents' house in Yorkshire. Books such as *The Pocket Guide to Turtles, Snakes and Other Reptiles* (2018) and *An Animal ABC* (2016) by Alice Pattullo (b.1988) anticipate her screenprints of White's tortoise *Timothy* (fig.185), swallows circling Selborne Church (fig.186), and *Gilbert* (fig.184), an imaginary scene of White writing letters in his study with a flock of birds looking in through the window.

Linocut was used by printmaker Christopher Brown (b.1953) to depict Timothy (fig.176) in an homage to his own childhood pet, and a reimagining of Nash's illustration *'It hobbles towards its benefactress'* (fig.175). Brown had been an assistant to the artist Edward Bawden and he recalled how, 'returning from Cornwall one year, Bawden suggested a detour to Selborne. He would have done a great job illustrating the book. He was extremely fond of both the Nash and Ravilious editions.'[5] In turn, Ravilious has also been a point of reference and admiration for Angie Lewin (b.1963), who created a wood-engraving of a feather-filled Victorian Sunderland mug featuring a vignette of a bird feeding its young in a nest (fig.183). Here, nature is unconstrained – the floral motifs burst from the container, as if the plants are untrammeled and there is no boundary between real nature and imaginary.

Michael Kirkman (b.1986) employed linocut for his print depicting the birds described by White as being 'the most unusual birds' he had ever observed, which frequented an ornamental piece of ground adjoining his garden: 'They used to march about in a stately manner, feeding in the walks, many times in the day; and seemed disposed to breed in my outlet; but were frightened and persecuted by idle boys, which would never let them be at rest.'[6] Kirkman's print (fig.172) features organic linear patterning of serpentine forms of tree, sunrays and feathers, only disrupted by the presence of one of the boys bursting into the scene. This subject had previously been tackled by several other artists, mostly notably Ravilious and Nash, and was also chosen by Mark Hearld (b.1974), whose collage *Hoopoes* (fig.179) focusses on the spectacle of these exotic-looking birds in the garden of The Wakes. Hearld is known for his collages and prints of birds in flight and small country animals such as squirrels, rabbits and mice, and he has illustrated books such as *A First Book of Nature* (2012) and *Nature Poems* (2019). He studied natural history illustration at the Royal College of Art under one of Bawden's pupils, John Norris Wood, and his view of nature is inherently English, rooted in the seasonal variations of spring, summer, autumn and winter, with an appreciation for folk art. When approaching White's writings, Hearld focussed on particular episodes as the stimulus for large paper collages in painted artist's frames. *Osprey* (fig. 178) relates to the description of a bird White had seen at Frensham Pond in Hampshire, also tackled by Ravilious, Clare Leighton and John Nash: 'It was sitting on the handle of a plough and devouring a fish: it used to precipitate itself into the water, and so take its prey by surprise.'[7] *Hare, Partridges and Pheasant* (fig. 177) illustrates a passage from White's fifth letter to Thomas Pennant in which he declared: 'The manor of Selborne, was it strictly looked after, with all its kindly aspects; and all its sloping coverts, would swarm with game; and even now hares, partridges, pheasants around; and in old days woodcock were a plentiful.'[8]

Ed Kluz (b. 1980) chose to depict the Hermitage (fig. 182) in homage to Grimm's oval frontispiece. He observed: 'The notion of the hermitage is one which perfectly reflects White's own relationship with nature; a raw dwelling where the boundaries between the human and natural world are frayed, allowing for quiet, deep contemplation and observation.' Kluz used the medium of scraperboard used by many illustrators in the 1930s and 1940s, a gesso panel with a black coating which is scratched off to reveal the white beneath. He noted: 'The ephemeral nature of the hermitage appealed to me and I chose to depict White's reimagined hermitage surrounded by a woodland animated by trees caught in the act of whipping and arching their branches. Here it is the natural world which is forceful and the hermitage a place of stillness.'[9]

The idea of place and place-making is also something explored by the wood-engraver Neil Bousfield (b. 1967) in his two wood-engraved diptychs *Gilbert's Tears I* and *II* (figs 173, 174). He has observed how, 'Place is an emotional construct made from memory, personal experiences and values projected on to the landscape spaces we occupy. I was interested in making work that explored, past, present, narrative and experience, and these are the ideas that underpin the work *Gilbert's Tears*'. He knew the landscape of Selborne from first-hand memory, having taken his dogs for walks along the routes and footpaths that Gilbert White had walked.

Like others before him Bousfield has explored the technique of wood-engraving and built on the traditional craft-based techniques pioneered by Thomas Bewick, but in a way that is not about nostalgia but that speaks equally to our times.

> Although the tools and principles Bewick would know, the materials and approach can only be considered contemporary. I enjoy the freedom of making my own blocks, scaling up the technique and working with multiple blocks and reduction methods to facilitate the use of colour. The method is slow, requires precise registration, and allows time to be spent developing each outcome.[10]

In his hybrid wood-engraving and linocut entitled *Likeness* (fig.187), Tristan Sherwood (b.1976) chose to explore the idea of Bewick's influence and the idea of mimicry in both art and nature through an egg housing a Bewickesque cuckoo, much larger than the scale of the diminutive original wood-engravings. He has commented:

> The many editions of *The Natural History of Selborne* have spawned numerous illustrative 'incarnations' of White's cuckoo. This scenario formed a poetic motif, through which I could illustrate the nature of similitude. Making work 'after' another artist could be seen as equivalent to the parasitic nature of the cuckoo. Thomas Bewick, Eric Ravilious and Agnes Miller Parker are clearly informed by their antecedents, unlike the cuckoo itself it is their distinct interpretations which nestle their own birds within our mind's eye. The discovery of so many versions of White's cuckoo was a further surprise and delight, each seemingly informing the latter. I placed the study in an egg to stress the idea of mimicry in art and nature and the oak grain print again mimics nature and references the tree the bird sits in.[11]

Sculptor and printmaker Jo Sweeting (b. 1965) created a woodcut print of a head exploring White's interest in the migratory patterns of birds: *Hybernaculum Head* (fig.192). White felt certain, as Aristotle and Pliny did, that some of the later born birds chose not to migrate and instead remained in nooks, sand tunnels and hollow trees, hibernating. Sweeting has commented how it, 'shows White observing birds in the landscape as an interior topography. White's head as a cerebrum of sky. The hirundines flight lines of remembered symmetry etched in his mind'.[12]

As it happened, 2020 was a very different year from what might have been expected, due to the outbreak of the coronavirus pandemic – lockdowns across Britain and the world meant that many of the events to mark the tercentenary of White's birth did not take place. However, what happened instead was that many people paused to observe nature on their doorsteps, to listen to the birds once the noise of the traffic had abated, and to value the importance of walks in the countryside. During the lockdown, Sweeting carved a partner to her *Hybernaculum Head* titled the *Isolare Head* (fig.190), showing a head filled with hirundines – house martins, swallows and swifts spinning in the sky. At the same time, she carved a series of limestone and sandstone pebbles with the names of birds on the 'Red List' of endangered British species, including the curlew, lark and sparrow (fig.191). Sweeting imagined these empty nests as fossils that might exist following the birds' extinction:

> As a pebble heads towards its own extinction, as it migrates and faces reduction through time, so does our bird population. The stone nests have the names of birds who are now endangered in our British habitat. The pebble nests stand to act as memorials to them unless action is taken to support their natural environments and to enable them to survive.[13]

As the artist Neil Bousfield has observed, 'If Gilbert White was here today, the Attenborough of his time, he would mourn the predicted loss of the bumble bee, birds, the impact of climate change, loss of habitat, biodiversity, and would be in tears for what is now only beginning to dawn upon us. Gilbert White is more relevant now than ever.'[14]

FIG.172
Detail from '*The most unusual birds*', 2020 (see fig.181).

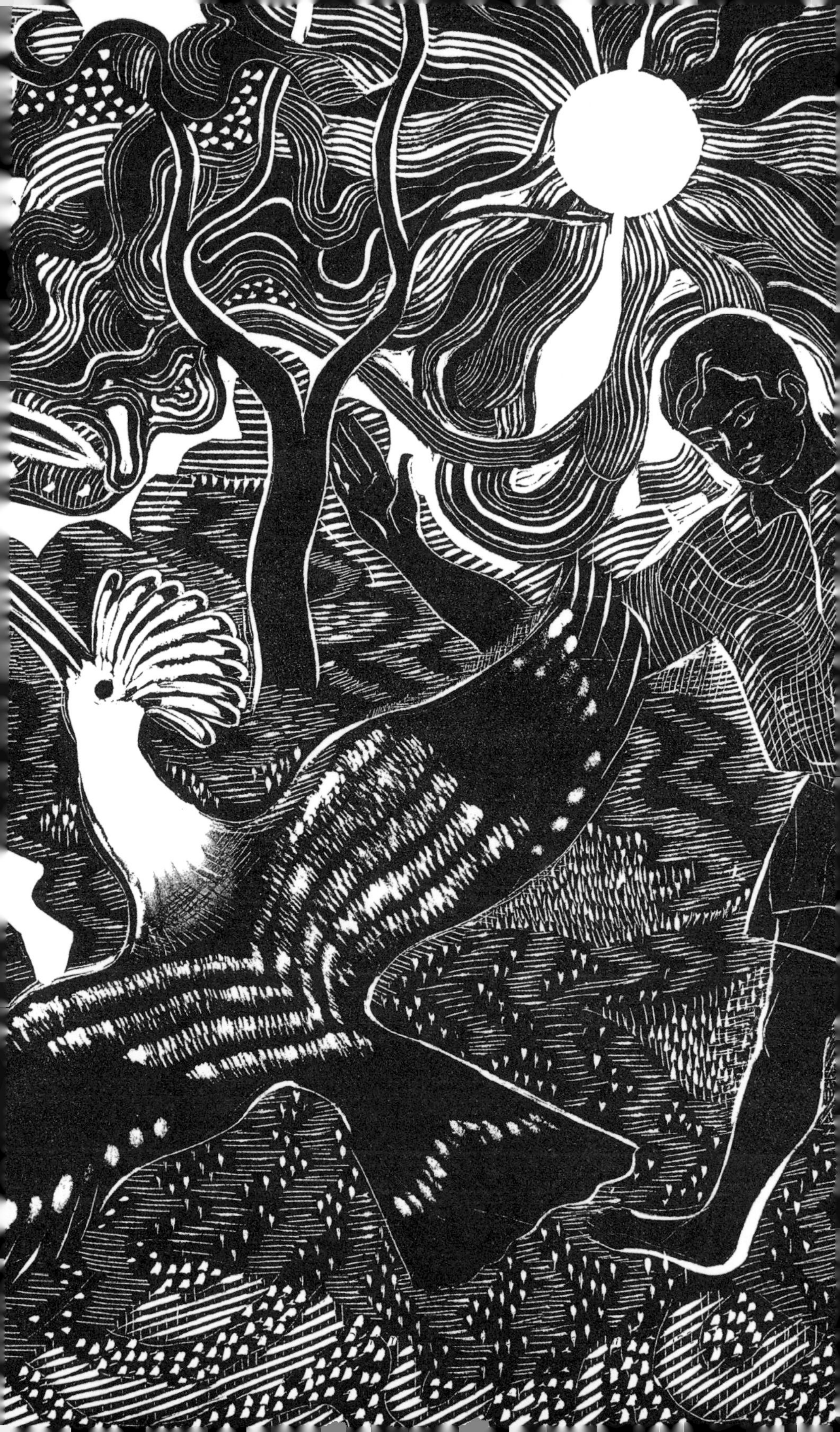

NEIL BOUSFIELD (b.1967)

FIG.173
Gilbert's Tears I, 2020.
Wood-engraving on paper (diptych).
Pallant House Gallery (commissioned from the artist, 2020).

FIG.174
Gilbert's Tears II, 2020 (detail).
Wood-engraving on paper (diptych).
Pallant House Gallery (commissioned from the artist, 2020).

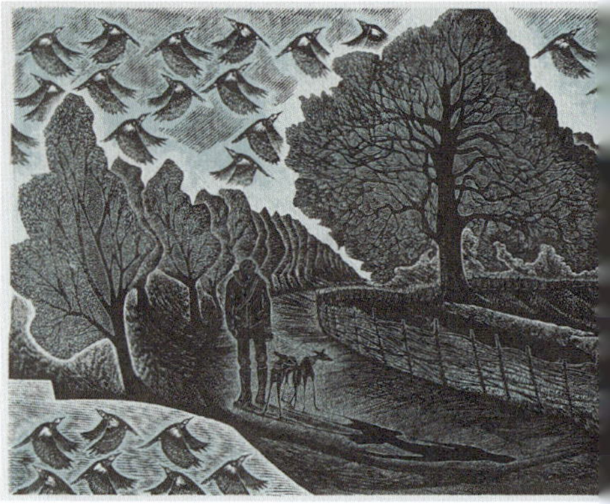

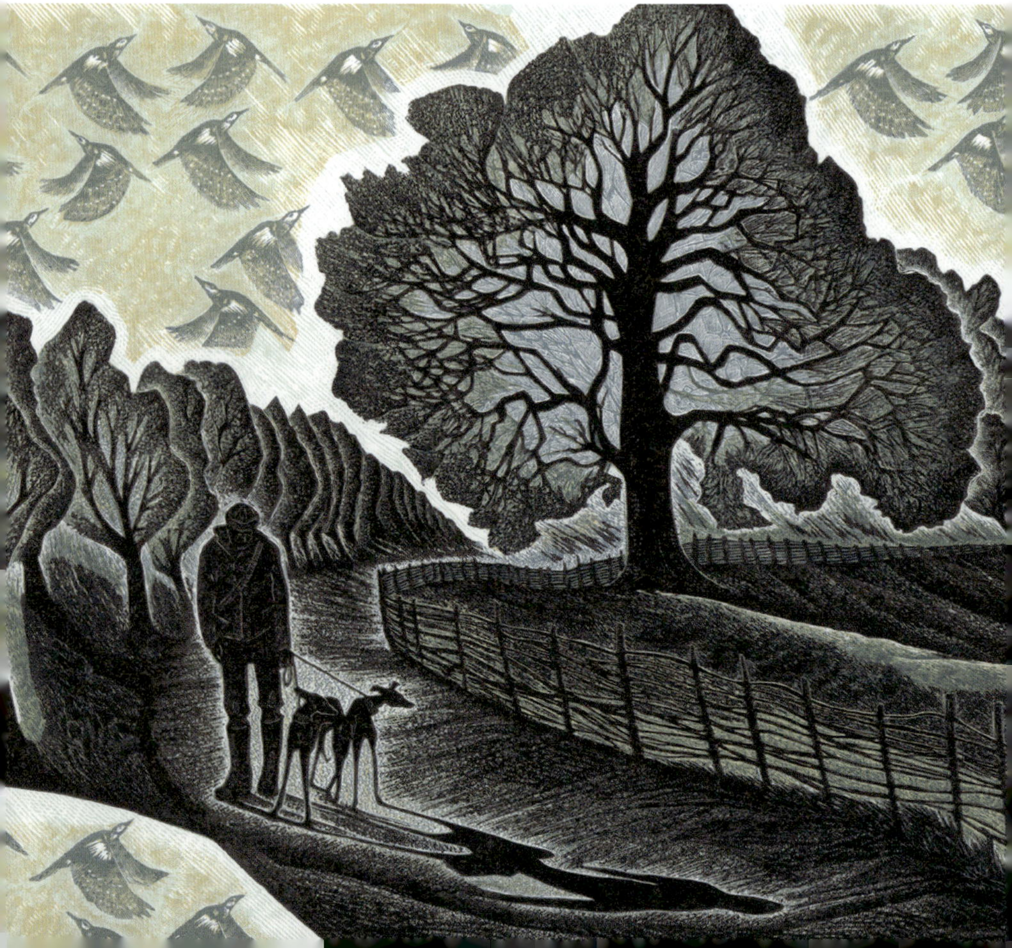

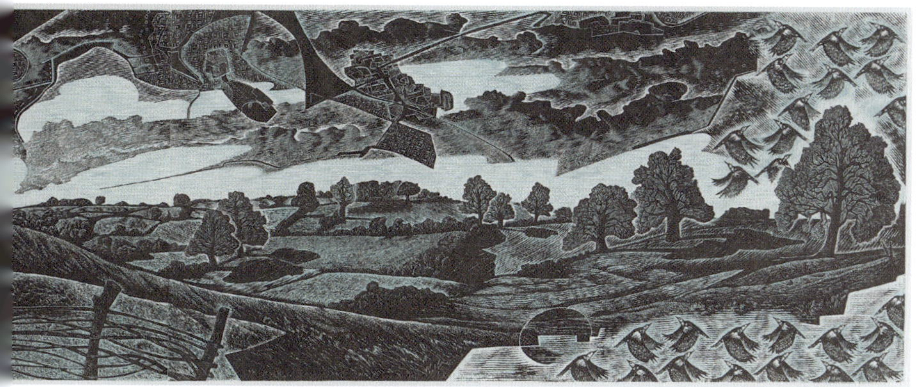

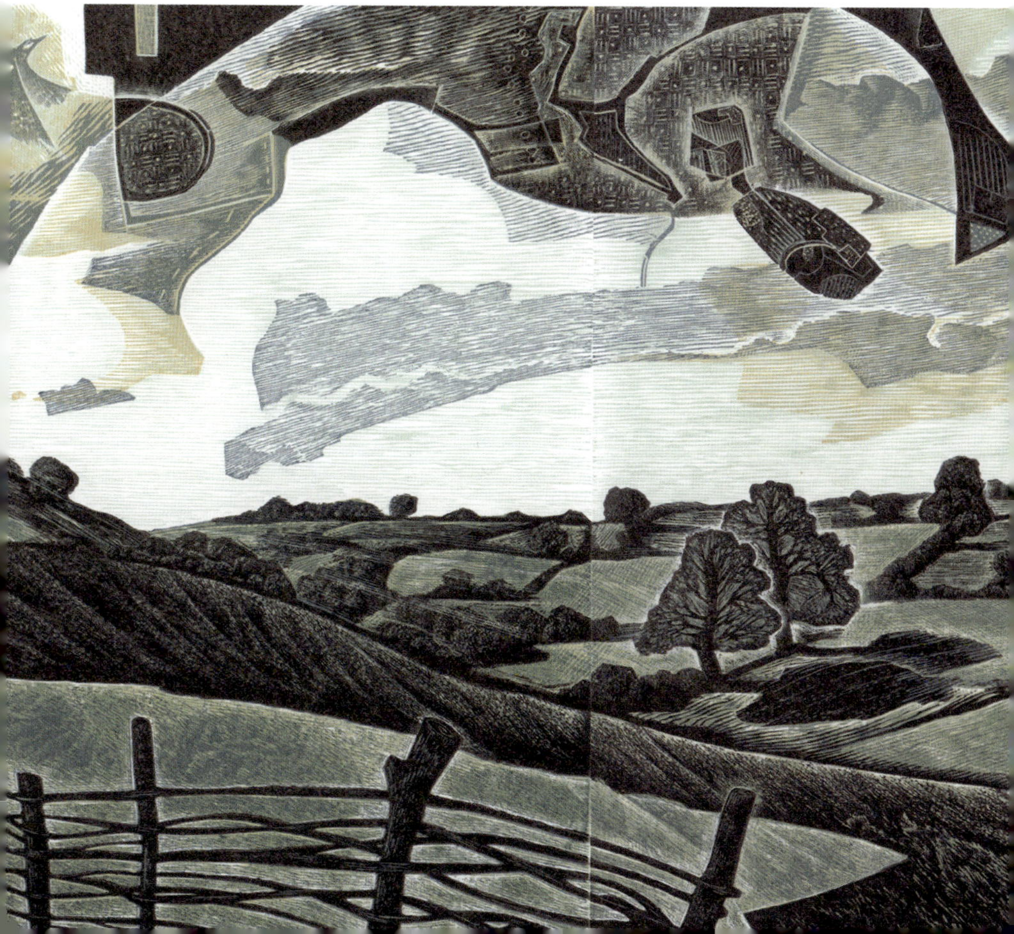

CHRISTOPHER BROWN (b.1953)

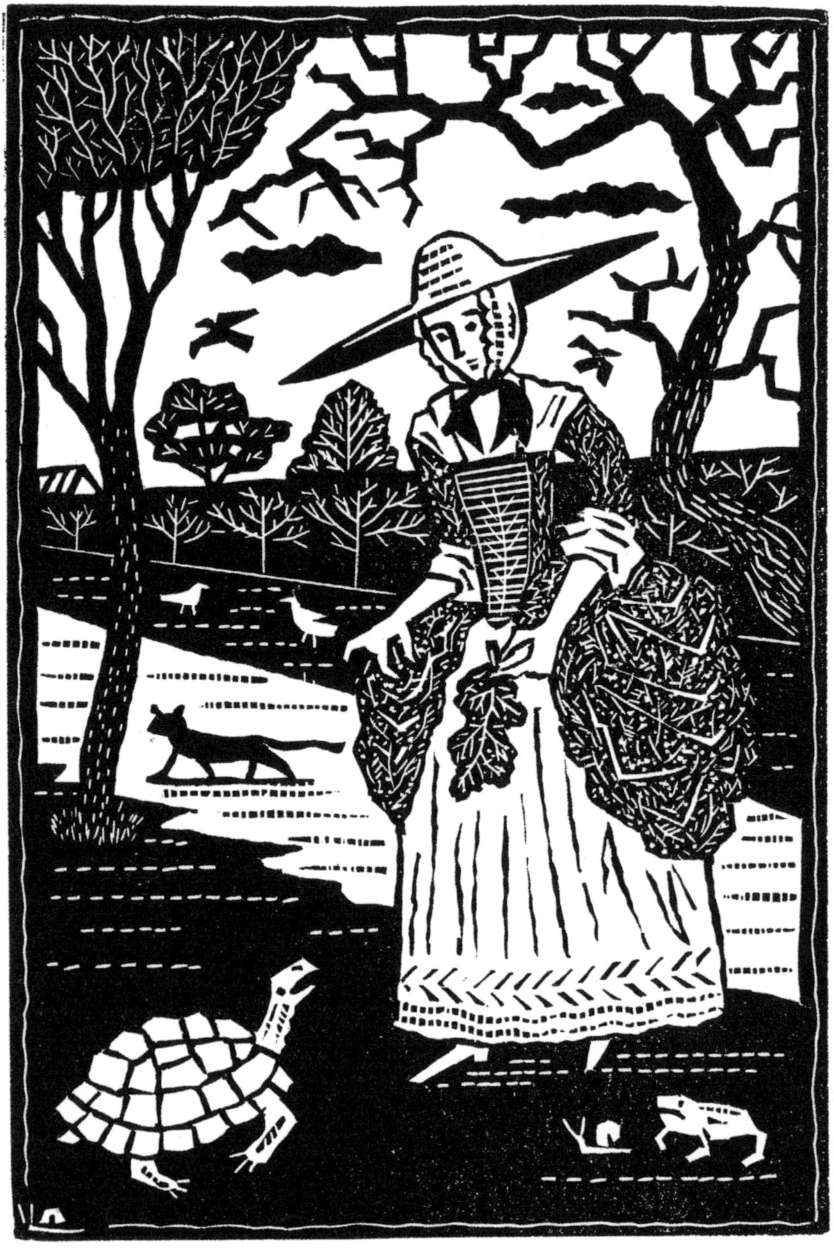

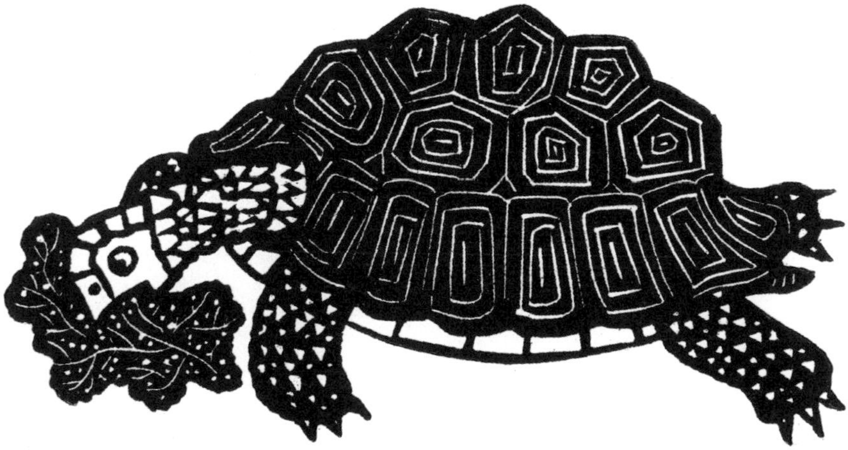

FIG.175 (opposite)
'It hobbles towards its benefactress', 2020.
Linocut on Zerkall paper. Pallant House Gallery (commissioned from the artist, 2020).

FIG.176
A Garden Pet, 2020.
Linocut on Zerkall paper. Pallant House Gallery (commissioned from the artist, 2020).

MARK HEARLD (b.1974)

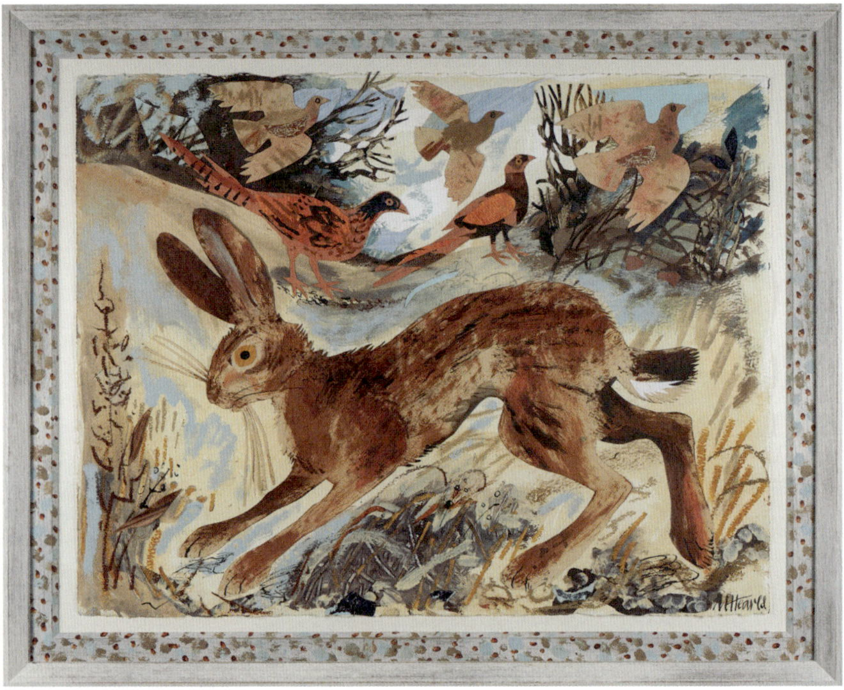

FIG.177
Hares, Partridges and Pheasant, 2020.
Collage and gouache on paper in artist's frame.
Private collection.

'The manor of Selborne, was it strictly looked after, with all its kindly aspects; and all its sloping coverts, would swarm with game; and even now hares, partridges, pheasants around; and in old days woodcock were a plentiful. There are few quails because they more affect open fields than enclosures; after harvest few landrails are seen.'
Letter V to Thomas Pennant, Gilbert White.

Fig.178 (opposite, above)
Osprey, 2020. Collage and gouache on paper in artist's frame. Private collection.

'It was sitting on the handle of a plough and devouring a fish: it used to precipitate itself into the water, and so take its prey by surprise.'

FIG.179 (opposite, below)
Hoopoes, 2020. Collage and gouache on paper in artist's frame. Pallant House Gallery (given by Robin Vousden, 2020).

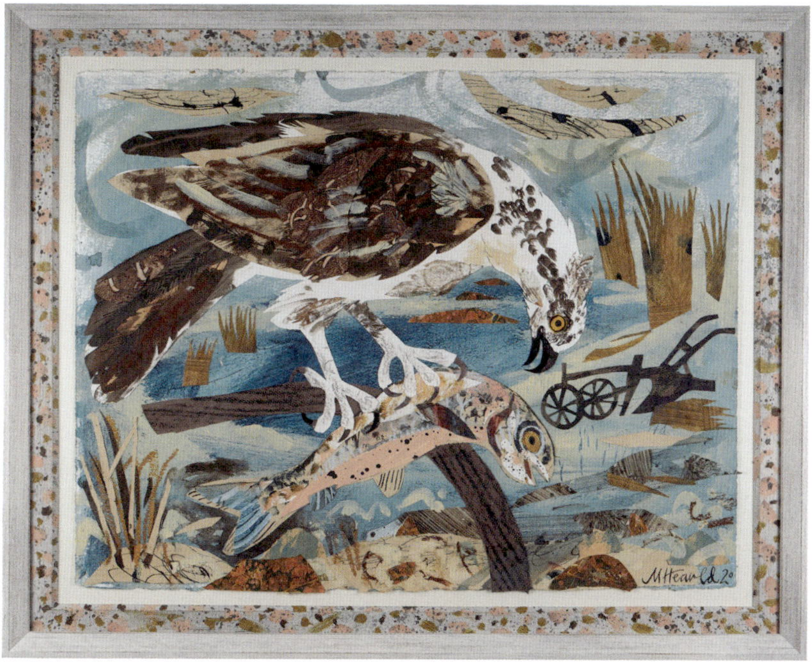

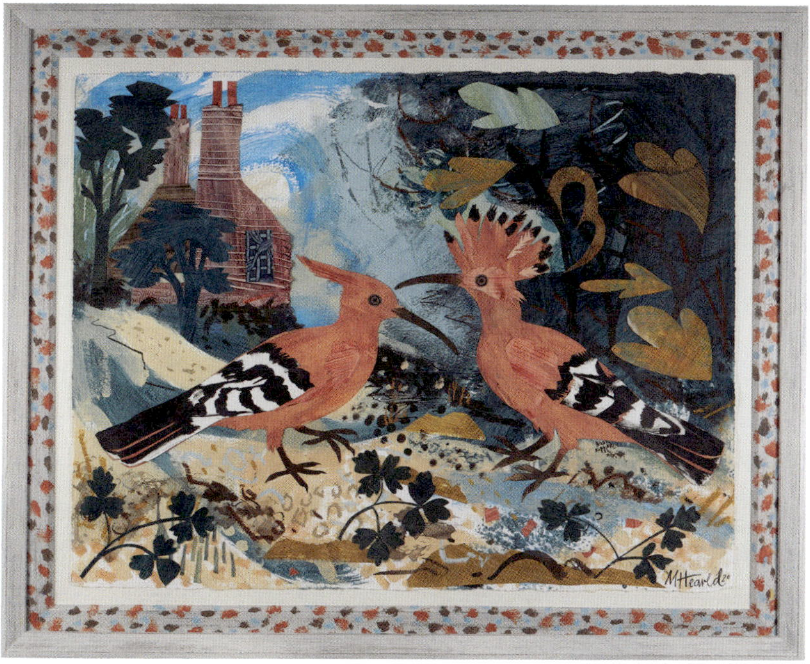

CLIVE HICKS-JENKINS (b.1951)

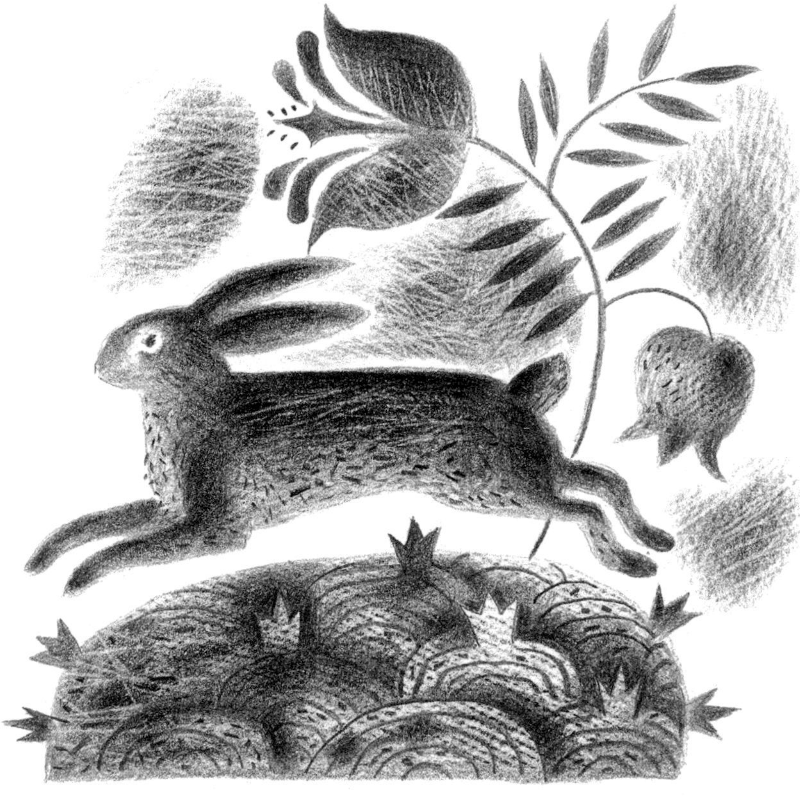

FIG.180
Hare, 2020. Pencil on paper. Pallant House Gallery (presented by the artist, 2020). This drawing was included as an illustration for *Charis in the World of* Wonders by Marly Youmans, published in 2020.

MICHAEL KIRKMAN (b.1986)

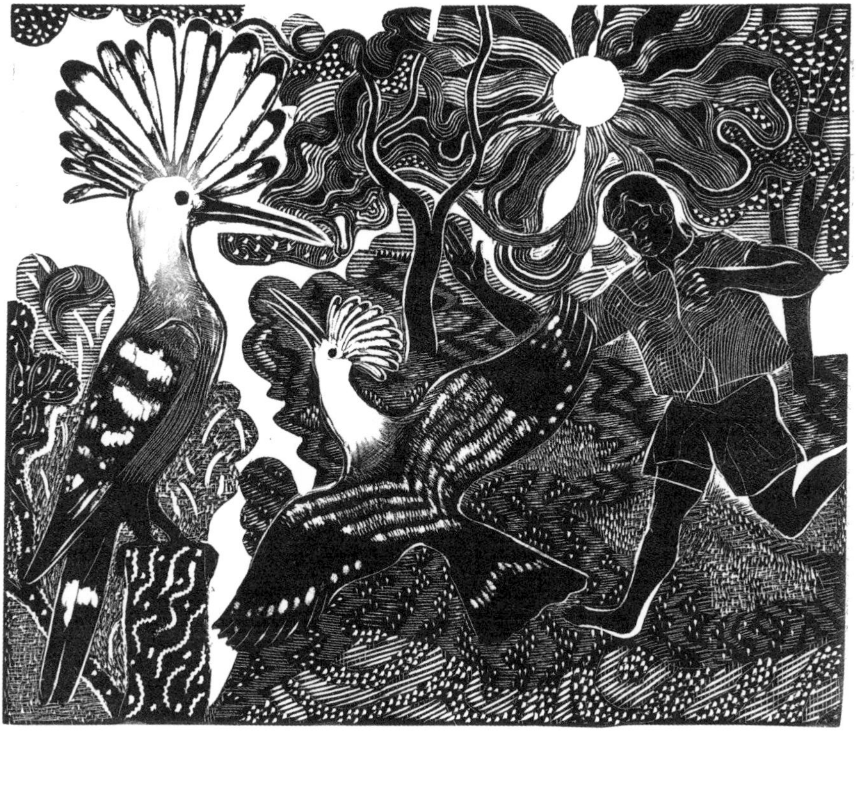

FIG.181
'The most unusual birds', 2020.
Linocut on paper. Pallant House Gallery
(presented by the artist, 2020).

ED KLUZ (b.1980)

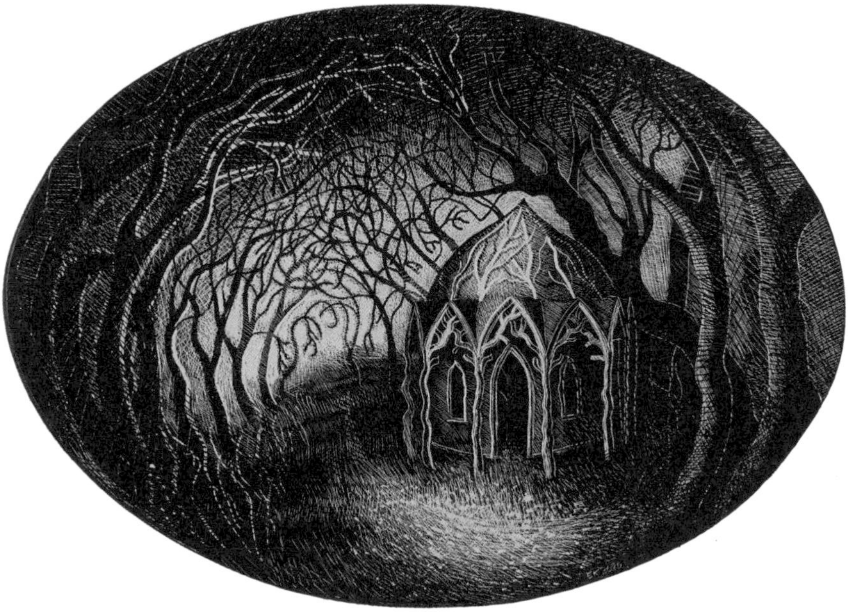

FIG.182
The Hermitage, 2020.
Scraperboard on paper.
Private collection

ANGIE LEWIN (b.1963)

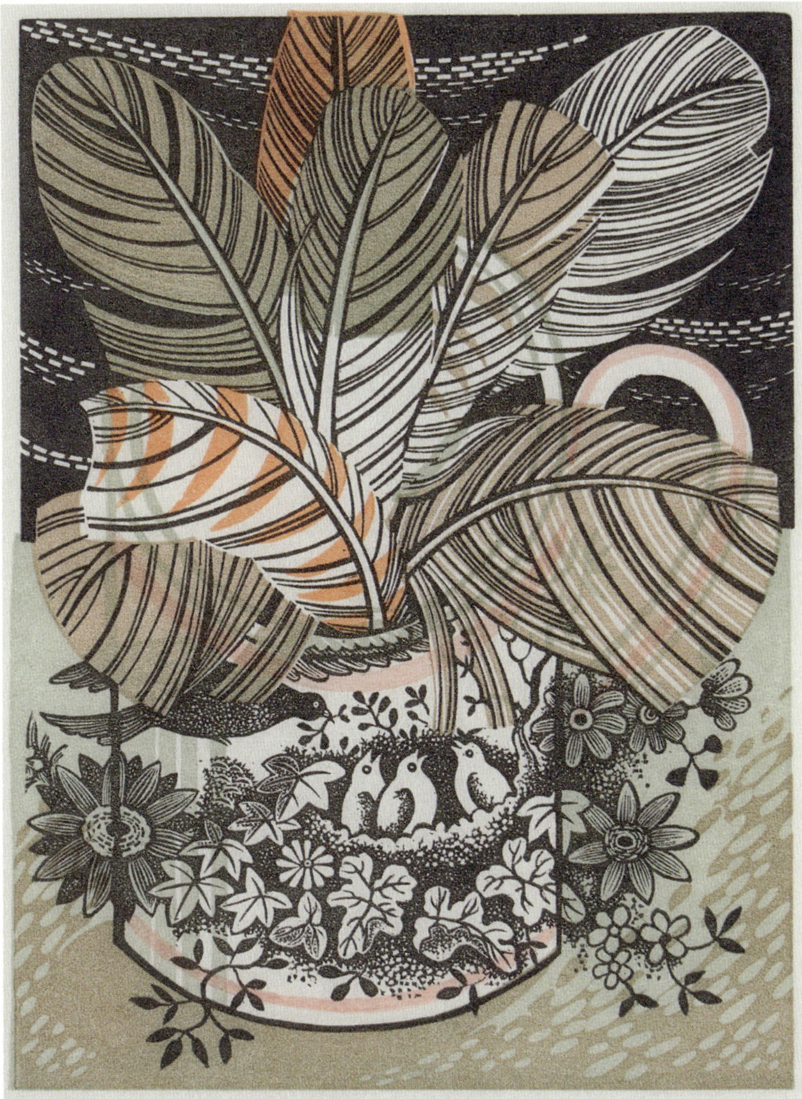

FIG. 183
Bird Cup, 2020. Wood-engraving on paper.
Pallant House Gallery (commissioned from the artist, 2020).

ALICE PATTULLO (b.1988)

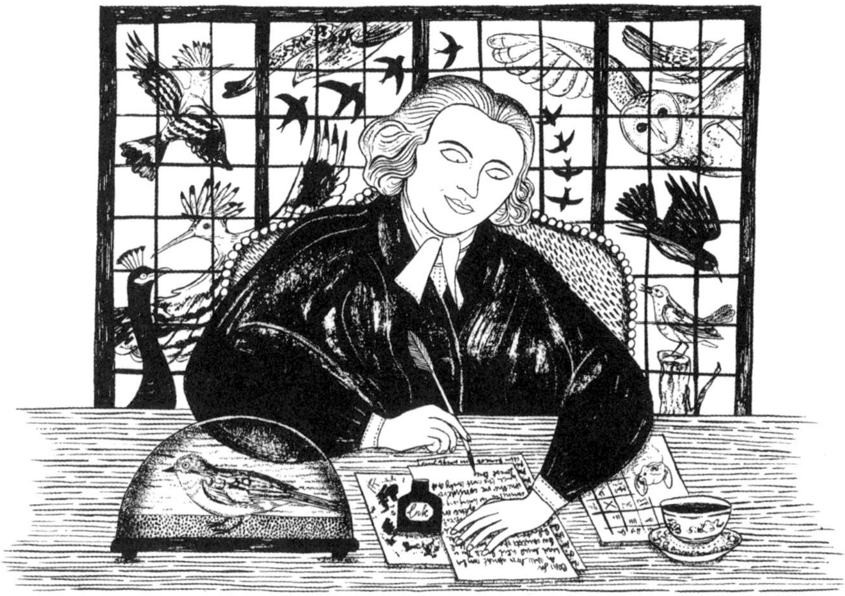

FIG.184 (above)
Gilbert, 2020. Screenprint
on paper. Pallant House Gallery
(commissioned from the artist, 2020).

FIG.185 (opposite, above)
Timothy, 2020. Screenprint
on paper. Pallant House Gallery
(commissioned from the artist, 2020).

FIG.186 (opposite, below)
Selborne Church, 2020. Screenprint
on paper. Pallant House Gallery
(commissioned from the artist, 2020).

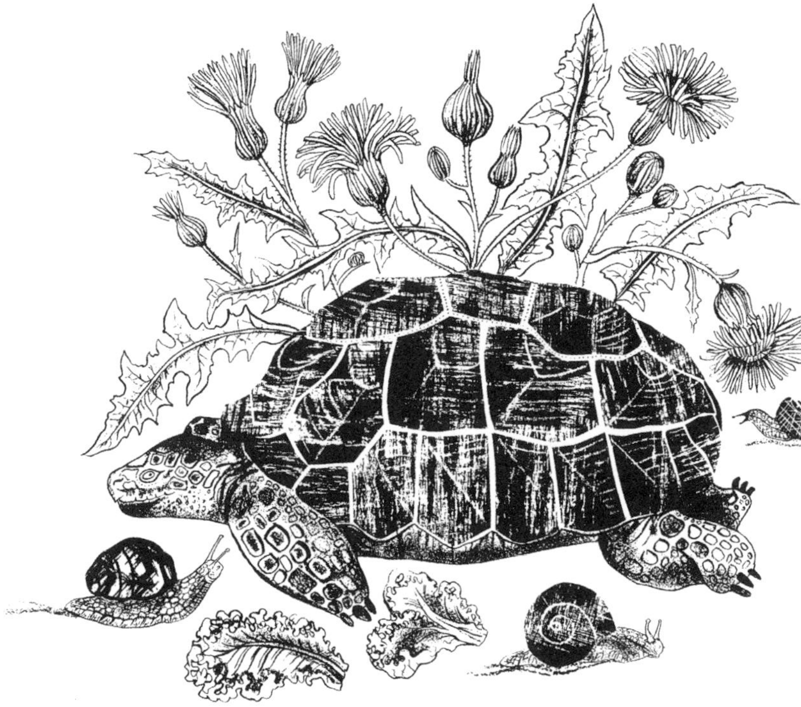

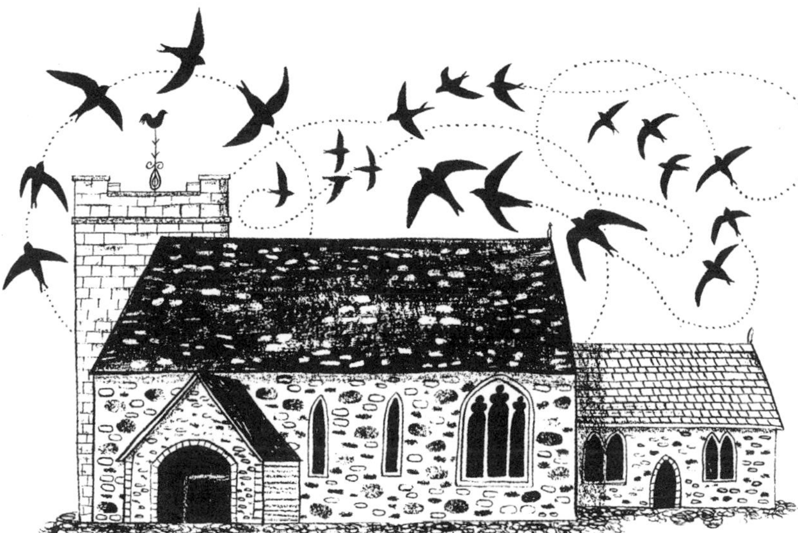

TRISTAN SHERWOOD (b.1976)

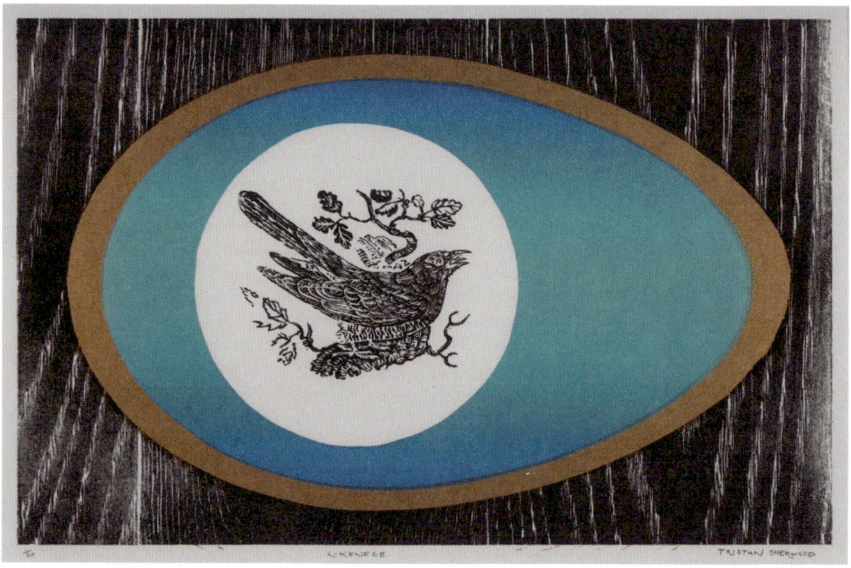

FIG.187
Likeness, 2020. Hybrid woodcut and linocut print on paper, edition of 20. Pallant House Gallery (commissioned from the artist, 2020).

EMILY SUTTON (b.1983)

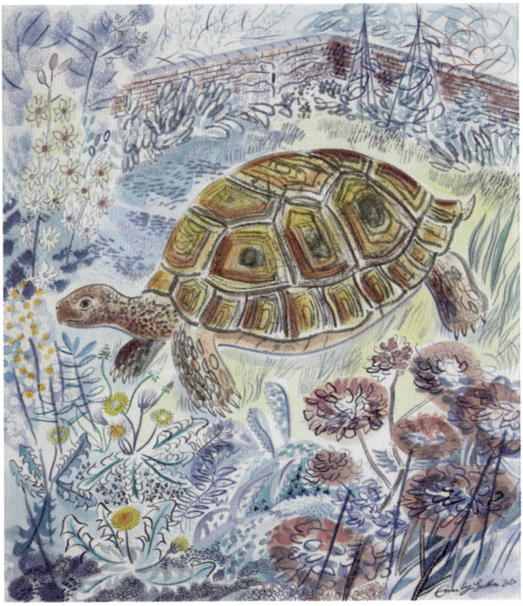

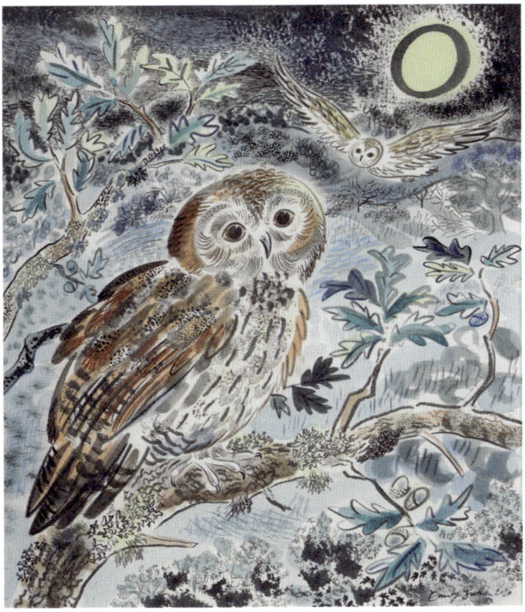

FIG.188
Tortoise, 2020. Watercolour and pencil on paper. Private collection.

FIG.189
White Owl, 2020. Watercolour and pencil on paper. Private collection.

JO SWEETING (b.1965)

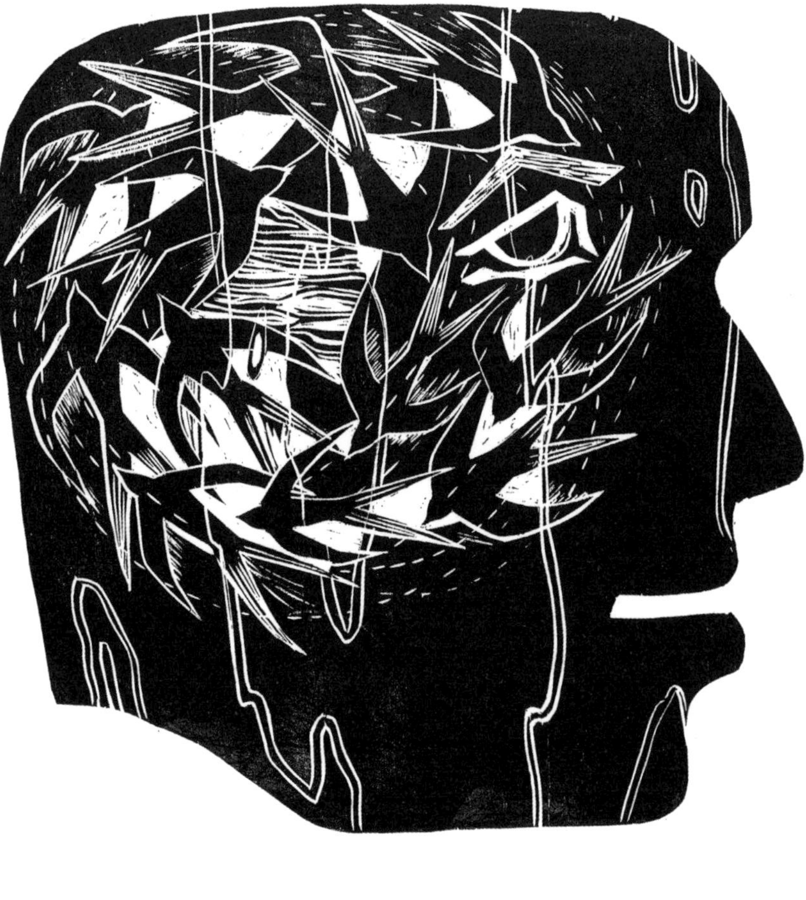

FIG.190
Isolare Head, 2020. Woodcut print on paper. Pallant House Gallery (commissioned from the artist, 2020).

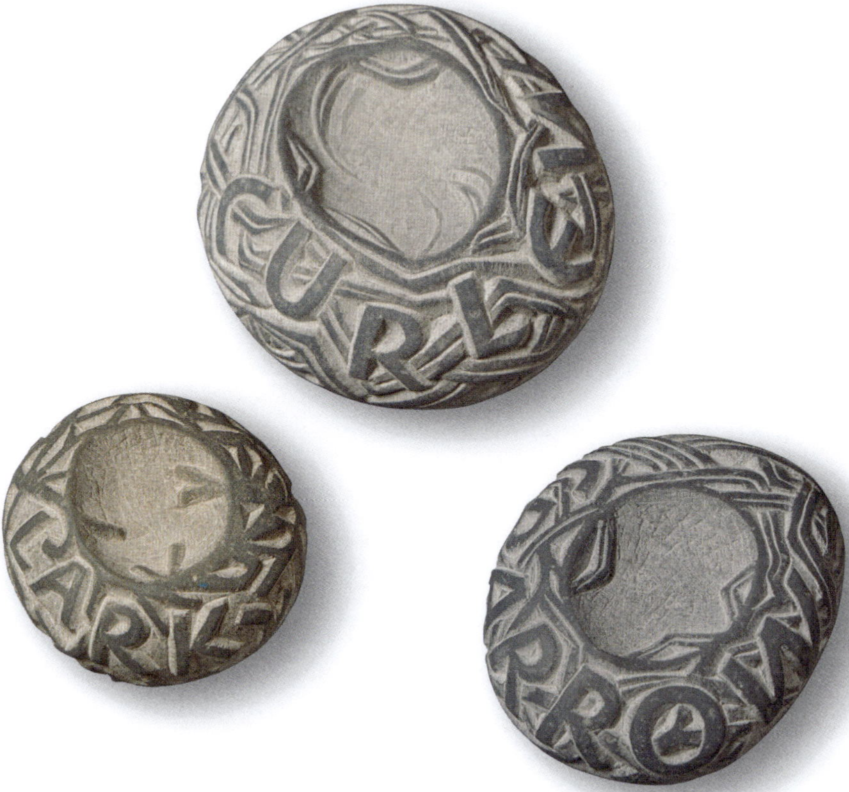

FIG.191
Empty Nests: Curlew, Lark and Sparrow, 2020.
Carved limestone and sandstone pebbles.
Private collection.

A VISION

FOR THE REVEREND GILBERT WHITE

Kathryn Bevis

The forest is alive today
and quick with wild devotion.
Bees hum, drunk on puffs
of pollen, censer-swung
from meadowsweet
and Queen Anne's lace.

Ferns stir themselves
to nod and bow; they sail
a summer breeze. Open-handed
to the sun, each pair of leaves
is a single prayer in a reef
of fractal-patterned green.

Damselflies flash and dart,
a fever of electric grace.
In the shade a foal gazes,
still as any seer;
her flanks are polished silver,
her tail an aspergillum.

The body of a world
at worship cries out
to be seen. The beech leaves
whisper in a psalm
to everything that flickers,
foams and gleams.

(Kathryn Bevis, Hampshire Poet 2020–1, was commissioned by Winchester Poetry Festival and Hampshire Cultural Trust to write this poem in celebration of the 300th anniversary of Gilbert White's birth.)

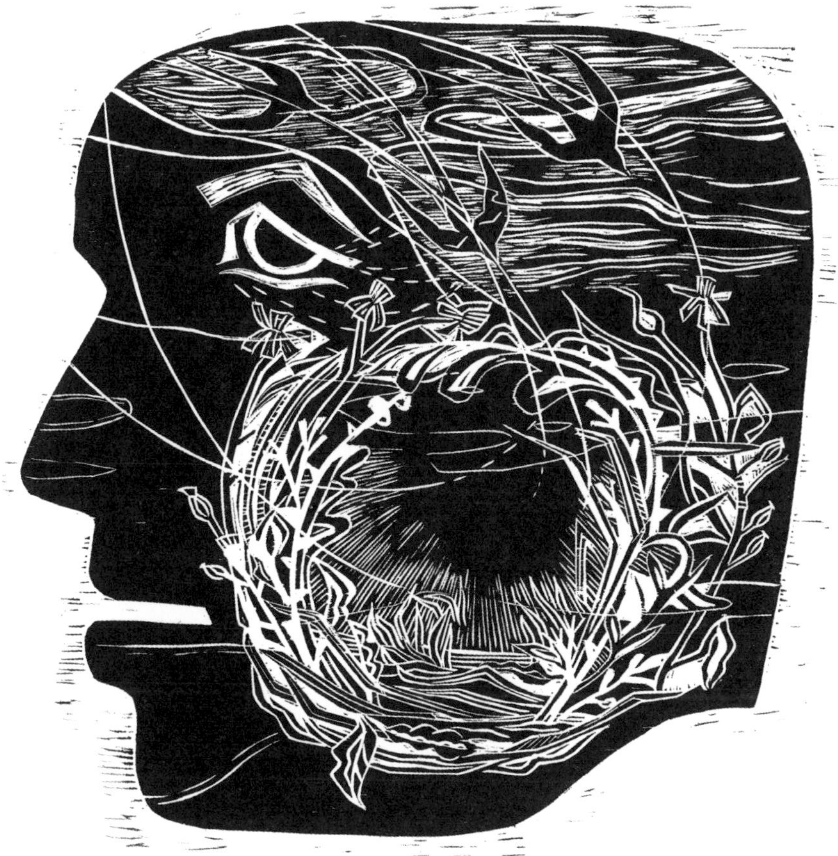

FIG. 192
Jo Sweeting, *Hybernaculum Head*, 2020.
Woodcut print on paper. Pallant House Gallery
(commissioned from the artist, 2020).

'A NIGHTINGALE FOR GILBERT WHITE: 5 APRIL 1768'

Jo Bell

Buds and shadows fatten, but the garden's lean.
A London smoke crawls west, and cucumbers
are tortoising across the sweat-sweet dung.

A nuthatch jars and clatters in the oak;
rooks get cocky in the Selborne copse. At last
the air is quick with bee-flies, kites and larks

and April falls across the parish like stained glass,
like rest for the broken-backed. The diarist
dashes off one word to stand for spring – *Luscinia!*

Colour blurs from every quickened hedge
into the woodsmoke hours. The nightingale
loops speechless syllables on every thorn.
Attention, after all, is prayer. Nothing goes unseen

(First published in *Kith* by Jo Bell, Nine Arches Press, 2015)

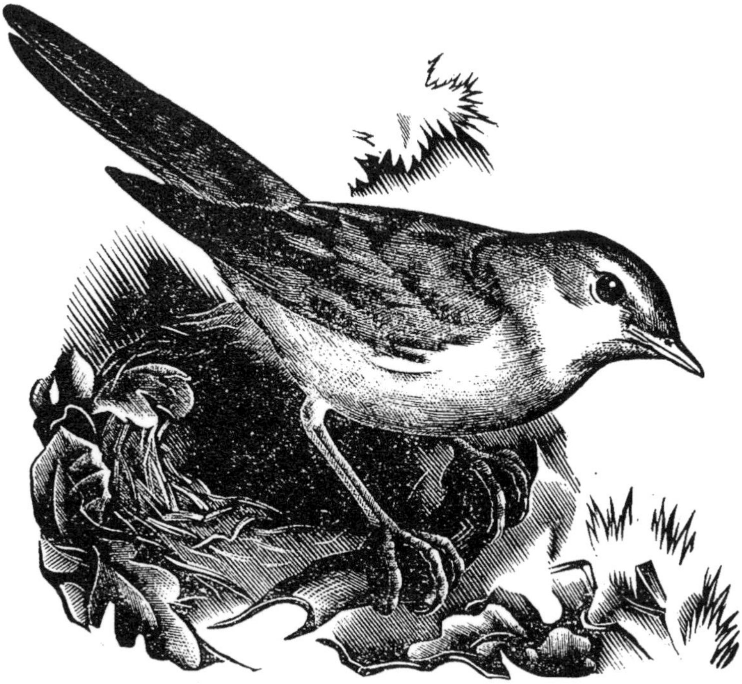

FIG.193
Agnes Miller Parker, *Nightingale*, 1941.
Wood-engraving on paper, in Leonard Russell, ed., *The Saturday Book 1941-2: A New Miscellany,* London: Hutchinson, 1941.

NOTES

OBSERVING NATURE: THE ILLUSTRATORS OF GILBERT WHITE'S NATURAL HISTORY OF SELBORNE

1. Although the publication date is 1789, *The Natural History of Selborne* actually came out in December 1788.
2. Gilbert White to the Hon. Daines Barrington, Letter I, Selborne, 30 June 1769.
3. Gilbert White to Thomas Pennant, Letter XXVIII, March 1770.
4. White to Barrington, Selborne, 12 April 1770.
5. It has been said that White's observations of earthworms so impressed Darwin that they formed the basis of his 1881 study *The Formation of Vegetable Mould Through the Action of Worms.*
6. Charles Darwin autobiography, quoted in Ernst Mayr, *The Growth of Biological Thought: Diversity, Evolution and Inheritance*, Harvard: Havard University Press, 1982, p.397.
7. Sir David Attenborough, Introduction to *The Gilbert White Museum Edition of The Natural History of Selborne by Gilbert White*, London: Shepheard-Walwyn, 1977, p.x (also on p.13 of this book).
8. White to Barrington Letter LVII, 1768.
9. White to Pennant, Letter XIII, Selborne, 22 January 1768.
10. White to Pennant, Letter XXXVI, September 1771.
11. Gilbert White, 21 January 1793.
12. James Harting in the Preface, September 1874, Gilbert White, *The Natural History and Antiquities of Selborne in the County of Southampton*, ed., James Harting, London: Bickers & Son, 1875, pp.ix-x.
13. White to Barrington, 20 May 1770.
14. Mabey 2006, p.83.
15. Virginia Woolf, 'White's Selborne', *New Statesman and Nation,* 30 September 1939, in Woolf 1966-67, p.122 (see pp.87–91 of this book).
16. Virginia Woolf, *ibid.*
17. Samuel Taylor Coleridge, quoted in Mabey 1993, p.6.
18. Gilbert White, 'Meteorological Observations', 22 November 1768.
19. Gilbert White, Letter XVII, December 1773.
20. White to Pennant, Letter XVI, Selborne, 18 April 1768.
21. Gilbert White, 'Meterological Observations: Thaw'. White read James Thompson, who he says: 'falls into fustian sometimes... though he thinks like a poet, is often faulty in his diction.' Quoted in Thomas 1917, p.112.
22. White to Barrington, Letter LIX.
23. Mabey 2006, *ibid.,* p.173.
24. W.T. William and G.H. Valllins, eds, *White's Natural History of Selborne*, London: Methuen, 1935, v, 8.
25. Auden 1973.
26. Warner 1946, p.7. More recently, American author Verlyn Klinkenborg wrote a novella from the sensibility of the tortoise: Klinkenborg 2006.
27. T.H. White, *The Once and Future King*, 1958, p.156.
28. Members of the Society of Antiquaries included Daines Barrington.
29. Constable to Archdeacon John Fisher, quoted in Mabey 2006, p.7.

30. Gilbert White, 26 October 1783.
31. For example, a very similar image of a 'Long Leg'd Plover' is one of a few birds listed in John Lightfoot's 1777 book *Flora Scotica* and was drawn by Moses Griffiths and engraved by Peter Mazell.
32. White, Letter LXIX, Selborne, May 1779.
33. Gilbert White to Rev. John White, Meonstoke, 9 August 1776.
34. See Mabey 2006, *ibid.,* p.165.
35. See Mabey 2006, *ibid,* p.166.
36. May 1776. See Rashleigh Holt-White, *Life and Letters of Gilbert White of Selborne, vol. I,* London: John Murray, 1901, p. 320.
37. Gilbert White to Rev. John White, Meonstoke, 9 August 1776.
38. Gilpin, *Essay on Prints*, 1768.
39. Gilbert White, 'The Invitation to Selborne', 1745. The quotation from the poem under the engraving refers to 'his straw-clad cell' but the actual wording of the poem is 'where the Hermit hangs the straw-clad cell'.
40. Gilbert White, *The Garden Kalendar*, 28 July 1763.
41. Mulso, 1 February 1785, from *Gilbert White: The Naturalist's Journals*, quoted in Mabey 2006, *ibid.* p.167.
42. White to Pennant, Letter III.
43. White to Pennant, Letter III.
44. White to Pennant, Letter III.
45. White to Pennant, Letter IX. The illustration features in Gilbert White, *The Natural History and Antiquities of Selborne, in the County of Southampton with Engravings, and an Appendix,* London: White, Cochrane & Co; Longman, Hurst, Rees, Orme, and Brown; J. Mawman; S. Bagster; J. and A. Arch; J. Hatchard; R. Baldwin; and T. Hamilton, 1813, p.484.
46. Gilbert White, *The Natural History of Selborne in the County of Selborne,* 1837 edition, with notes by Edward Turner Bennett, Secretary of the Zoological Society, p.435.
47. Edward Turner Bennett, *ibid.,* p.435.
48. Thomas Bewick, *The History of British Birds,* Newcastle: Beilby & Bewick, 1797.
49. Gilbert White includes a 'List of the Summer Birds of Passage which I have discovered in this neighbourhood, ranged somewhat in the order in which they appear' in his first letter to the Hon. Daines Barrington, written 30 June 1769.
50. Piper 1942, p.7.
51. See Spalding 2009, p.125.
52. See James Harting, ed., *ibid.*
53. White to Pennant, Letter II.
54. White to Pennant, Letter XXXV, Selborne, 1771.
55. James Russell Lowell, *Latest Literary Essays and Addresses,* 1891, quoted in Mabey 2006, *ibid.,* p.8.
56. These were two drawings by White's friend Thomas Chapman, drawn in a copy of Alexander Pope's *The Iliad* now in the British Library.
57. Richard Kearton, 'Introduction' to Gilbert White, *The Natural History of Selborne with notes by Richard Kearton*, London: Cassell, 1902.
58. Kearton, *ibid.*
59. Kearton, *ibid.* p.viii.
60. 'Gilbert White' in Thomas 1917, p.109.
61. Alexander Gascoyne and Horace Hincks were based in Nottinghamshire, as was Joseph Whitaker, and so this may be why these stained-glass artists were selected.
62. In a letter dated 23 November 1961 to Commander R.H.S. Rodger of Selborne, in the church records, Horace Hincks stated: 'I painted the window under Mr. Gascoyne's supervision, there is a possibility that a list (of the birds) may be in the possession of Mr. Whittaker or one of his family. The colouring of the birds was taken from two volumes of 'British Birds', a work by a naturalist author and artist whose name and name of his publishers, again, unfortunately fails me. Probably these volumes were Mr. Whittaker's and would be returned to him on completion of the work.'
63. See Angela Bartie, Linda Fleming, Mark Freeman, Tom Hulme, Alex Hutton, Paul Readman, 'Pageant of Selborne: Gilbert White's Village', *The Redress of the Past,* http://www.historicalpageants.ac.uk/pageants/1185/. The authors note that it is uncanny that Virginia Woolf's last novel, *Between the Acts* (1941), which is about the performance of a pageant staged by a female Pageant Master, should have been written so shortly after her 1939 essay on *White's Selborne* and within a few years of the 1938 staging of the Selborne Pageant. Sei Kosugi has asked: 'Why did this symbolic encounter between Mary Kelly and Virginia Woolf occur at a specific moment of history

(the late 1930s) and at the Hampshire village of Selborne?' See Sei Kosugi, 'Representing Nation and Nature: Woolf, Kelly, White', in *Locating Woolf: The Politics of Space and Place*, ed. A. Snaith and M. Whitworth, Basingstoke, 2007, p.89. The novelist Susan Barrett, in her novel *Making a Difference,* has conjectured that Woolf herself visited the pageant in 1926, though no evidence for this exists. See Susan Barrett, *Making a Difference: A Novel*, Victoria, Canada, 2006, p.297.

64. *Pageant of Selborne*, copy of advertising flier, Hampshire Archives and Local Studies, Winchester. TOP273/3/38.
65. Keith Shakleton's poster for the Shell Guide to Hampshire, painted in 1958, features a copy of White's *Natural History* propped up on a windowsill before a landscape view populated with birds and butterflies.
66. Leighton 1932, pp 94-95.
67. Gilbert White, *The Natural History of Selborne,* ed., E.M. Nicholson, London: Thornton Butterworth Ltd., 1929.
68. Quoted in Jane Hill, *The Sculpture of Gertrude Hermes*, London: Lund Humphries, 2011, p.16.
69. Brett 1993, p.16.
70. Hermes to McCance, 10 June 1932, quoted in Brett, *ibid.,* p.16.
71. William McCance letter to Hermes, 3 May 1932, quoted in Brett, *ibid.,* p.16.
72. Harris 2010, p.171.
73. Brett, *ibid.,* p.16.
74. Hermes to Burdon Evans at the Gregynog Press, 27 March 1932, quoted by James Hamilton, 'Gertrude Hermes at Gregynog' in Condry 1988, p.26.
75. Gregynog Press Directors' meeting 15 November 1932, cited by Hamilton in Condry, *ibid.,* p.29.
76. See Condry, *ibid.* Each of the prints had been published in editions of 12 in the 1930s. A further edition of 20 of each print was published in 1988 with paper impressed with the Greynog Press die, separate from the edition bound into the book.
77. *The Spectator,* 2 May 1931, p.20. See Walter Johnson (ed.), *The Journals of Gilbert White*, London: G. Routledge & Sons Limited, 1931. Johnson had previously written *Gilbert White: Pioneer, Poet and Stylist*, London: John Murray, 1928.
78. Gilbert White, *Writings of Gilbert White of Selborne,* ed., H.J. Massingham, 2 vols., London: The Nonesuch Press, 1938.
79. H.J. Massingham, *ibid.,* p.xxvi.
80. Harry Carter to Eric Ravilious, May 1937.
81. Eric Ravilious to Diana Low, 15 March 1939, quoted in Greenwood 2008, p.256.
82. White to Pennant, Letter 22, Selborne, 2 January 1769.
83. Inspired by visits to Firle House in Sussex, Ravilious had created several watercolours of greenhouses in around 1935.
84. Diana Sainsbury-Green, letter to Helen Binyon, 15 April 1973, East Sussex Records Office, ACC 9534.
85. Eric Ravilious to Helen Binyon, 30 January 1936, quoted in Greenwood, *ibid.,* p.254.
86. White to Barrington, Letter 62, *c.*1787.
87. White to Barrington, Letter 61, *c.*1787.
88. White to Barrington, Letter XLII, Selborne, 7 August 1778.
89. Eric Ravilious to Diana Low and Clissold Tuely, Christmas Day 1937, quoted in Greenwood, *ibid.,* p.254.
90. Eric Ravilious to Helen Binyon, 7 November 1937, quoted in Greenwood, *ibid.,* p.254. Ravilious' original engraved woodblocks are lost. The 1938 book was printed by the Nonesuch Press from electrotypes: a chemical method for forming metal parts that exactly reproduce the original engraved wood block. It was the standard method for reproducing plates for letterpress printing from the late 1800s onwards, through which many of the editions of Selborne would have been printed. In 1999, the typographer and book designer Peter Guy (1938–2009) donated five of these electrotype blocks to the Fleece Press with the following note: 'When I was a very junior lad at the Bodley Head – almost forty years ago! – I had the chore of throwing out a stack of old books, all of which were total junk; except that among them were these electros of a few of Ravilious' engravings for *A Natural History of Selborne*.' Originally used as the headpiece to volume one, and tailpiece to volume two of the Nonesuch Press edition, these were printed by the Fleece Press in a special edition, including the *Memorial to Gilbert White* (fig. 5), based on his simple tombstone in the churchyard with just the letters GW. See Simon Lawrence, *When I was a Lad at the Bodley Head:*

Ravilious blocks rescued by Peter Guy, Upper Denby: The Fleece Press, 2016.

91. White to Pennant, Letter XI, Selborne, 9 September 1767.
92. Recounted by Harry Carter to Eric Ravilious, 12 January 1938, quoted in Greenwood, *ibid.,* p.254.
93. White, Letter XLIII, Selborne, 9 September 1778.
94. Leonard Russell, ed., *The Saturday Book 1941-2: A New Miscellany,* London: Hutchinson, 1941. The same issue also features a journal-style article by H.J. Massingham, who had edited the 1938 Nonesuch Press edition – this was used for the texts to accompany Miller Parker's images, which suggests that he may have been involved with her commission. These were reprinted from the original blocks now in the National Library of Scotland in Rogerson 1990, pp 32-33.
95. Gilbert White, *The Natural History of Selborne Illustrated by Clare Leighton,* ed., James Fisher, Harmondsworth: Penguin, 1941.
96. James Fisher, *ibid.* p.x.
97. White also wrote a poem entitled 'A Harvest Scene' (see p.22) which describes how alongside the reaper 'attendant by his side/ His faithful wife, sole partner of his cares / Bears on her breast the sleeping babe; behind / With steps unequal, trips her infant train: Thrice happy pair, in love and labour join'd.'
98. White to Barrington, Letter XXVII, 12 December 1775.
99. Gilbert White, *The Natural History of Selborne Illustrated by Clare Oldham*, ed., James Fisher, London: Cresset Press, 1947. This includes an introductory essay, incorporating material from 'A New Judgement on Gilbert White' broadcast in 1945.
100. Bertram Nicholls, 'A Self-Taught Wood-Engraver: Claire Oldham' in *The Studio,* January 1944, vol.127, no.610, pp 18-20. I am grateful to Sarah Gilroy and Gill Clarke for bringing this article to my attention.
101. Nicholls, *Ibid.*
102. White to Pennant, Letter 12, 4 November 1767.
103. White to Pennant, Letter 16, 18 April 1768.
104. White to Barrington, Letter 24, 15 August 1775.
105. White to Barrington, Letter VII, Ringmer, near Lewes, 8 October 1770.
106. Lewis 1947.
107. John Lewis in 'Foreword' to Gilbert White, *The Natural History of Selborne with Drawings by John Nash*, ed., John Lewis, London: Lutterworth Press, London, 1951, p.5.
108. Colvin 1986, p.55. Colvin provides a detailed account of Nash's work on the Lutterworth Press edition, including correspondence with John Lewis and details of Nash's concerns about the layout and production of the book.
109. White to Barrington, Letter LVII, undated.
110. White to Barrington, Letter XVIII, Selborne, 29 January 1774.
111. Nash to Clarence Elliott, quoted in Colvin, *ibid.,* p.54.
112. Colvin, *ibid.,* p.55.
113. White to Pennant, Letter 17, Selborne, 18 June 1768.
114. White to Barrington, Letter 13, 12 April 1772.
115. White to Barrington, Letter 15, Selborne, 8 July 1773.
116. John Lewis, *ibid.,* p.5.
117. Gilbert White, *The Natural History of Selborne,* Ipswich: Limited Editions Club, 1972. The hand-coloured proofs for these are now in the Gilbert White & the Oates Collections in Selborne.
118. Nash to Griselda Lewis, undated, August 1971, quoted in Colvin, *Ibid.,* p.70.
119. Interestingly, while Ravilious and Leighton's illustrations for the scene in their respective editions of White's *Selborne* are quite different, Leighton's print is remarkably close to an earlier Ravilious wood-engraving entitled *Boy Birdsnesting.* It was first exhibited in an edition of 15 at the Eighth Annual Exhibition of the Society of Wood Engravers, and so it seems likely that Leighton must have known this print.
120. *Extracts from the Natural Histories of Selborne by Gilbert White Illustrated with wood engravings by Mrs H.J. Faulkner and W.V. Rose*, Birmingham: City of Birmingham School of Printing, 1950. The pamphlet was printed under the direction of Leonard Jay at the Birmingham School of Printing, a department of the College of Arts and Crafts. The compositors' work was done by boys in the Junior School; it was printed by students attending the Letterpress Machine classes; and the wood-engravings were by Mrs H.J. Faulkner and W.V. Rose, students at the Harborne Branch School of Art.

121. Gilbert White Museum, Ibid.
122. Gilbert White, *The Natural History of Selborne: A new edition with engravings,* Richard Mabey, ed., London, Melbourne, Auckland, Johannesburg: Century, 1988.
123. Gilbert White, *The Natural History of Selborne,* ed., W.S. Scott, London; Folio Society, 1961. These were printed by Cotswold Collotype Co. Ltd. in Wotton-under-Edge, and the book was printed and bound by Richard Clay & Co Ltd, Bungay.
124. Joanna Selborne in Clinton and Selborne 1989.
125. Jenkins 1999, p.254.
126. Gilbert White, 'Advertisement to the First Edition', Selborne, 1 January 1788.

CONTEMPORARY RESPONSES TO GILBERT WHITE

1. Holding the exhibition in Chichester, some 26 miles from Selborne, was appropriate not only for the association with many of the Modern British artists who were part of the history, but because of local connections mentioned by White in his letters – his uncle had purchased Timothy from a Chichester sailor, his sister lived nearby at Chilgrove, and White had visited the nearby Goodwood Estate to view the aforementioned moose.
2. Clinton, *ibid.* 1989.
3. 'Perfect Distillations', review of *Drawn to Nature* at Pallant House Gallery, *Sunday Telegraph*, 18 April 2020.
4. Marly Youmans, *Charis in the World of Wonders*, San Francisco: Ignatius, 2020.
5. Christopher Brown to the author, February 2020.
6. Gilbert White, Letter XI, Selborne, 9 September 1767.
7. Gilbert White to Thomas Pennant, Letter XXXIX, Selborne, 9 November 1773.
8. Gilbert White, Letter V to Thomas Pennant.
9. Ed Kluz, *Pallant House Gallery Magazine*, Summer 2020.
10. Neil Bousfield, *Pallant House Gallery Magazine*, Summer 2020.
11. Tristan Sherwood, *Pallant House Gallery Magazine*, Summer 2020.
12. Jo Sweeting to the author, February 2020.
13. Jo Sweeting to the author, August 2020.
14. Neil Bousfield, *Pallant House Gallery Magazine*, Summer 2020.

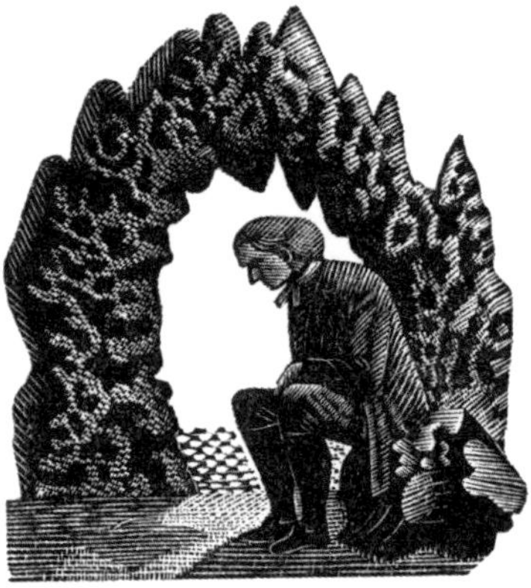

SELECT BIBLIOGRAPHY

The list of books below relates to relevant books on White, the artists and the themes in this book. As there are over 400 editions and reissues of Gilbert White's *Natural History of Selborne* these are not listed in full here, but a very helpful and compressive list of editions between 1789 and 1983 was compiled by Ian F. Lyle A.L.A Librarian, Royal College of Surgeons, and is included in Mabey 1988.

W.H. Auden, 'Posthumous Letter to Gilbert White', *The New York Review,* 18 October 1973.

Alun Jones, Deborah, *The Wry Romance of the Literary Rectory*, London: Thames & Hudson, 2013.

Blunden, Edmund, *English Villages*, London: William Collins, 1941.

Brett, Simon, *Gertrude Hermes: The Wood-Engravings*, Aldershot: Scolar Press/ Ashgate, 1993.

Clinton, Patricia and Selborne, Joanna, *The Natural History and Antiquities of Selborne: An Exhibition to celebrate the Bicentenary of the Publication of Gilbert White's Book*, Manchester: Manchester Polytechnic Library, 1989.

Colvin, Clare, *John Nash Book Designs*, Colchester: The Minories, 1986.

Condry, William, *Wood Engravings by Gertrude Hermes being Illustrations to Selborne with Extracts from Gilbert White,* Newtown: Gwasg Gregynog, 1988.

Friend, Andy, *Ravilious & Co. The Pattern of Friendship*, London: Thames & Hudson, 2017.

Gordon, Seton, *Wild Birds in Britain*, London: B.T. Batford, 1938, with illustrations by John Nash.

Greenwood, Jeremy, *Ravilious: Engravings*, Woodbridge: Wood Lea Press, 2008.

Harris, Alexandra, *Romantic Moderns: English Writers, Artists and the Imagination from Virginia Woolf to John Piper*, London: Thames & Hudson, 2010.

Harris, Alexandra, *Weatherland: Writers & Artists Under English Skies*, London: Thames & Hudson: 2015.

Hewit, John, 'The "Nature" and "Art" of Shell Advertising in the Early 1930s,' in *Journal of Design History,* vol. 5, no. 2, 1992, Oxford University Press, pp 121–39.

Jenkins, Simon, *England's Thousand Best Churches*, London: Penguin Books, 1999.

Klinkenborg, Verlyn, *Timothy; or, Notes of an Abject Reptile*, New York: Alfred A. Knopf, 2006.

Lambirth, Andrew, *John Nash: Artist & Countryman*, Norwich: Unicorn Press, 2019.

Leighton, Clare, *Wood-Engraving and Woodcuts*, London: The Studio, 1932.

Lewis, John, *A Handbook of Printing Types with Notes on the Style of Composition and Graphic Processes used by Cowells*, Ipswich: W.S. Cowell Ltd, 1947.

Mabey, Richard, *Gilbert White: A Biography of the Author of The Natural History of Selborne,* London: Profile Books: 1986.

Mabey, Richard (ed.), *Gilbert White: The Natural History of Selborne*, London: Century, 1988.

Mabey, Richard, *Gilbert White*. London: JM Dent, 1993.

Piper, John, *British Romantic Artists*, London: William Collins, 1942.

Powers, Alan, *Eric Ravilious: Artist & Designer*, London: Lund Humphries, 2013.

Rogerson, Ian, *Agnes Miller Parker: Wood-engraver and Book Illustrator, 1895–1980*, Wakefield: Fleece Press, 1990.

Russell, James, *Ravilious Wood Engravings*, Norwich: The Mainstone Press, 2013.

Scott, Walter S., *White of Selborne,* London: Falcon Press, 1950.

Spalding, Frances, *John Piper, Myfanwy Piper,* Oxford: Oxford University Press, 2009.

Thomas, Edward, A *Literary Pilgrim in England,* London: Methuen, 1917.

Townsend Warner, Sylvia, *The Portrait of a Tortoise Extracted from the Journals and Letter of Gilbert White,* London: Chatto & Windus, 1946.

Uglow, Jenny, *Nature's Engraver: A Life of Thomas Bewick*, London: Faber & Faber, 2006.

Unden, Grant, *Understanding Book Collecting*, Woodbridge: Antique Collectors Club, 1982.

Woolf, Leonard, ed., *Virginia Woolf: Collected Essays III,* London: Chatto & Windus, 1966–67.

PICTURE CREDITS

IMAGE CREDITS

Pallant House Gallery would like to thank all those who have given permission for images to be reproduced and all those who have supplied images, including private collectors and sources who prefer to remain anonymous. Every effort has been made to trace copyright holders and to obtain permission for the use of copyright material; the Gallery would like to apologise for any unintentional errors or omissions in the list below, and would like to be notified of any corrections that should be made in future reprints or editions.

© Neil Bousfield: figs. 173, 174
© Estate of G. Stanislaus Brien: fig.30
© Christopher Brown: figs. 175, 176
© Estate of Eric Fitch Daglish: figs. 32–34, 77–82
© Estate of Eric George: fig.31
© Estate of Irene Hawkins: fig.12
© Mark Hearld: figs. 177–179, cover and endpapers
© Estate of Gertrude Hermes: figs. 35–38, 83–86
© Clive Hicks-Jenkins: fig.180
© Michael Kirkman: figs. 172, 181
© Ed Kluz: fig.182
© Estate of Lynton Lamb: fig.64
© Estate of Clare Leighton: figs. 45–46, 115–123
© Angie Lewin: fig.183
© Estate of Frederick Marns / Shepheard-Walwyn (Publishers) Limited: figs. 6, 170
© Agnes Miller Parker Estate: figs. 43 a–c, 110–114, 193
© Alice Pattullo: figs. 184–186
© John Nash Estate: figs. 3, 51–58, 60, 73, 140–158
© Estate of Claire Oldham: figs. 1, 47–50, 70, 124–139
© The Piper Estate / DACS 2021: figs. 66, 159–162
© Howard Phipps: fig.65
© Jude Tarrant: figs. 69a–c
© Tristan Sherwood: fig.187
© Estate of Charles Stewart: figs. 62, 63
© Emily Sutton: figs. 171, 188–89
© Jo Sweeting: figs. 190–192
© Chris Wormell: figs. 67–68, 163–168, 169a–b

PHOTOGRAPHY CREDITS

Courtesy Jeremy Greenwood/The Wood Lea Press: figs. 104–109
Barney Hindle (photographer): figs. 2, 14–17, 19–20, 23–24, 32–38, 44–50, 52, 56, 60–61, 64, 66–68, 70, 76–86, 102–103, 110–111, 115–139, 143–144, 159, 163–168, 169a–b, 171–172, 175–176, 180–191
Liss Llewellyn Fine Art: figs. 102–103
Courtesy Little Toller Books: figs. 5, 10, 39–42, 71–72, 87–101, 194–196
Simon Martin (photographer): frontispiece, figs. 3–4, 7– 9, 11–13, 18, 21a–g, 22, 27– 29 and details), 43a–c, 51, 53–55, 57–59, 62–63, 65, 73–74, 75a–c, 112– 114, 140–142, 145–158, 160–162, 193
Courtesy Shell Heritage Art Collection: figs. 30–31
Courtesy Jude Tarrant: figs. 69a–c

DECORATIVE MOTIFS

Eric Ravilious, illustrations to *The Writings of Gilbert White of Selborne*, H.J. Massingham, ed., London: Nonesuch Press, 1938.

p.1: Tailpiece, volume two.
p.5: *Contents*, headpiece in volumes one and two.
p.95: Tailpiece, volume one.
p.184: Headpiece, volume two, 'The Antiquities of Selborne'.
p.188: Tailpiece, volume two (a 'striped verte longue')
p.192: *The Astronomer*, endpiece, volume one.

INITIAL LETTERS

Edmund Hort New, illustrations to *The Natural History of Selborne by Gilbert White*, Grant Allen, ed., London: The Bodley Head, 1900.

pp 9, 25, 87, 153

AUTHOR'S ACKNOWLEDGEMENTS

When interviewing for new positions, I often set the candidates the task of speaking about their favourite work of art. It reveals aptitudes and interests that other questions somehow never uncover. A couple of years ago, when recruiting for a Personal Assistant, I was delighted to hear one of the interviewees speak with insight and enthusiasm about Eric Ravilious' *Hoopoes* from *The Writings of Gilbert White*. It was an engaging presentation, and set me thinking about White. I took my own copy of the book down from the shelf to re-read it and, not long afterwards, I visited the Gilbert White and Oates Collection where Pallant House Gallery had lent items of furniture over many years. I fell in love with Selborne and happened to notice that White's tercentenary was approaching in the following year. Seeking out illustrated editions of the *Natural History* artists, I discovered more and more. Just when I thought I'd found them all, I would discover another and then another (and there are others not included in this book). Of course, the candidate who had presented Ravilious' *Hoopoes* got the job and has been a wonderful assistant. I would therefore like to thank Becky Lyle for setting me on this journey. What I hadn't appreciated until much later was that Becky's father, Ian F. Lyle, Librarian of the Royal College of Surgeons, had been a huge fan and collector of Gilbert White material and had compiled the excellent list of all the publications in 1988, which was so helpful in tracing editions of the book.

When I started asking artists whether they would like to contribute a new illustration for the *Natural History* I was pleasantly surprised that they all enthusiastically agreed. Clive Hicks-Jenkins had recently completed drawings in the spirit of White and Bewick, while the others all created new works. I would therefore like to thank the artists who have created such wonderful works: Neil Bousfield, Christopher Brown, Mark Hearld, Clive Hicks-Jenkins, Michael Kirkman, Ed Kluz, Angie Lewin, Alice Pattullo, Tristan Sherwood, Emily Sutton and Jo Sweeting. In addition, I am grateful to Christopher Wormell and Howard Phipps, who had illustrated earlier editions, and all the artists' estates who have given permission for works to be reproduced, and to Jo Bell and Kathryn Bevis for permitting the reproduction of their poems written to mark the tercentenary.

Several individuals and institutions lent works to the exhibition and kindly permitted their reproduction here. I would like to acknowledge Simon Lawrence of the Fleece Press; Sara Cooper, Joe Hill and Karen Taylor of the Towner Gallery in Eastbourne; Joanna Selborne and her late husband John (the Earl and Countess of Selborne); Graham Williams of the Florin Press; Hugh Fowler Wright; David Leighton of the Clare Leighton Estate; Paul Liss of Liss Llewelyn; Kimberley James of the Gilbert White and Oates Collections, and Khalid Aziz, the Chair of Trustees. Robin Vousden not only lent to the exhibition, but acquired the Mark Hearld *Hoopoes* and generously donated it to Pallant House Gallery's permanent collection.

Conducting research during a pandemic with libraries closed was not without its challenges, and I am grateful to the following for their assistance and support: Adrian and Gracie Cooper, Stephen Daglish, Emyr Evans at the National Library of Wales, Clara Farmer at Chatto Books, Andy Friend, Sarah Gilroy and Gill Clarke, Sharon Hambly, Simon Hilton, Christina Janssen of the Scottish Gallery, Gillian Lamb, Rebecca Lyons and Helen Valentine at the Royal Academy, Richard Mabey, Alejandro Martínez Rodríguez, Aidan McNeil at Art UK, Jane Mote and Josh Kershaw of Beech to Beach who produced a memorable short film and Hugh Bonneville who narrated some of White's writings, Jackie and David Russell, Tim Russell, Alice Strang, Phil Sweeting, Jenny Uglow, Robert Updegraffe at St Mary's Church in Selborne (who enabled me to access the stained glass when the church was closed) and Claudia Wolfers. I would particularly like to thank John Gill, Jeremy Greenwood, Alexandra Harris, Nick Higbee and Alan Swerdlow for reading and commenting on my texts, and for their support and encouragement both for the exhibition and book.

I am immensely grateful to Sir David Attenborough for permitting us to reprint his essay on White, which remains relevant over forty years after it was first written; to Adrian Hunt for his beautiful design; Miranda Harrison for her careful editing of the texts; Jane Horton for her diligent work on the index; and to Mark Hearld for the wonderful cover. I would like to thank all of my colleagues and Trustees at Pallant House Gallery, but particularly Nick Benham, Eleanor Chant, Russell Le Page, Sarah Norris, Leo Powell, Tess Viljoen, Louise Weller and our Head of Publications Harriet Olsen, and Mark Eastment and his colleagues at Yale University Press.

This book would not have been possible without the generous support of Stuart Southall, the John Booth Charitable Trust, and a circle of Subscribers:

Janet Aidin
Peter and Alison Allen
Ruth Barrett-Danes
Chris and Lindsay Cawte
Alan Coombe and Lucy Thorpe
Ronald Davidson-Houston
Mr and Mrs E.D. Dickson
Ann and Doug Cleaver
Mark Eastment and Angela Prysor-Jones
Dr Chris Ford
Julie Giles
Guy Haslam
Lindsay Hughes
Cathy and Edward Hutton
Maggie Keeble
Jamie and Julia Korner
Sarah Lambie
Alison Paton and Hugh Lane
Deborah Paxford
Di Pickering
Howard and Judy Potter
Stephen Rayment
Andrew Shore
Greg and Katherine Slay
Mary Sore
Ian Neville George Stenning
Judith R. Thompson
Hugh and Clare Twiss
Brigitte and Alan Wasdell
Michael Wright

And all who wish to remain anonymous.

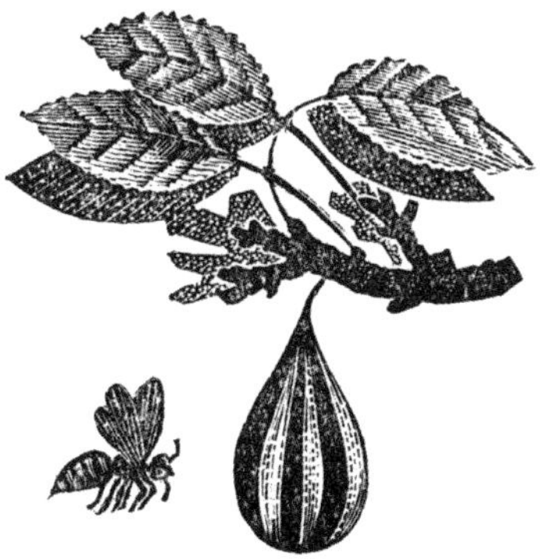

INDEX

NOTE: Page numbers in italics refer to illustrations.

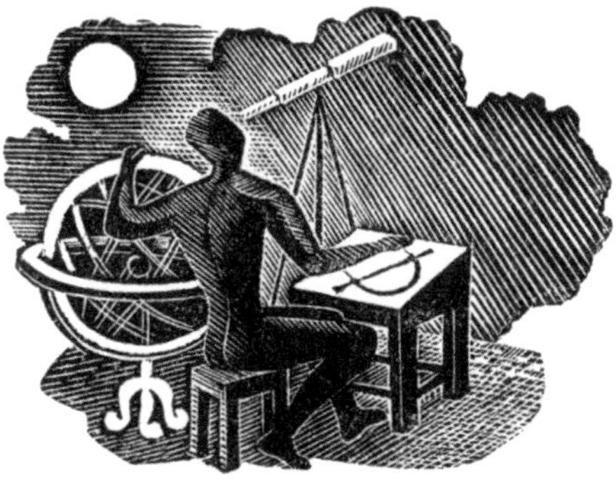

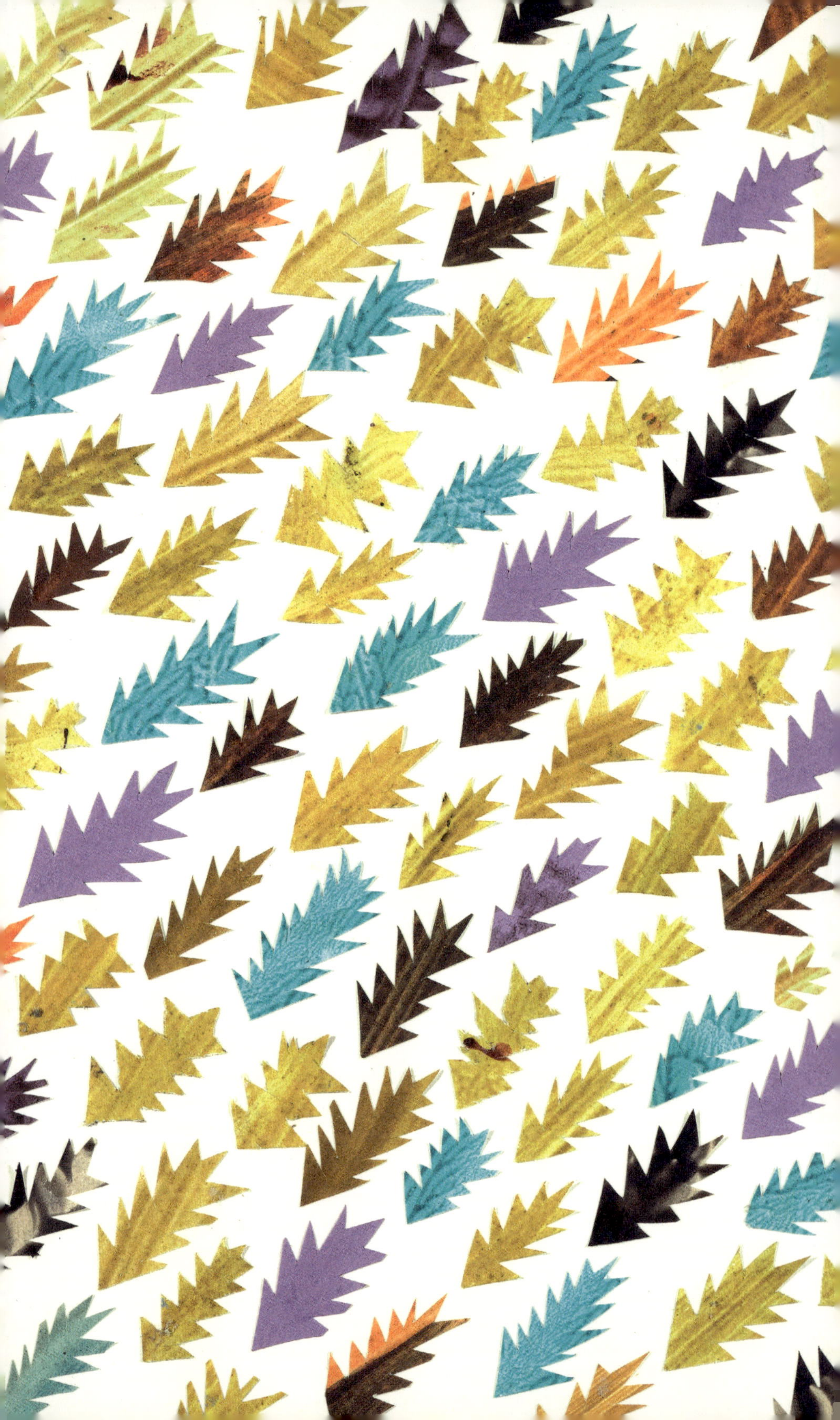